Livestock Rearing in the Tropics

Ian MacDonald and John Low

MACMILLA

First published 1985 by
MACMILLAN EDUCATION LTD
London and Oxford
Companies and representatives throughout the world

ISBN 0–333–39242–6

13 12 11 10 9 8 7 6 5 4
09 08 07 06 05 04 03 02 01 00

This book is printed on paper suitable for recycling and made from fully managed and sustained forest sources.

Printed in Hong Kong

A catalogue record for this book is available from the British Library.

Contents

Preface

This book has been written as a reference book for people advising farmers, for farmers themselves and for agricultural students.

In many countries there are excellent sources of information on livestock. Often, however, this information is highly technical and is neither available nor particularly suitable. It is hoped that this book will help to fill the gap and provide readers with practical information for planning and managing livestock enterprises. It attempts to show that different ecological areas require different breeds of animals and different systems of management. In addition, the smaller producer may require simpler management practices to fit his marketing situation and his available capital.

Many agricultural departments are now realising that animal husbandry and arable cropping must become much more integrated and that farm advisers cannot be only crop specialists. The emphasis on cereal production over the last 30 years is now leading, in many countries, to adequate supplies and a lowering of cereal prices. We need therefore to move towards mixed farming and this may include farm forestry, orchard crops, fish farming and animal husbandry.

We are grateful to the veterinary staff of the Wellcome Foundation in Kenya for their advice on health management.

Ian MacDonald
John Low

1 Cattle

Tropical and exotic breeds

Cattle represent a valuable asset in both traditional and modern agriculture. They provide meat, milk, skins and draught power. Additionally they may, in a traditional society, be an essential part of the social system, representing a family's wealth; or they may be regarded as a survival kit by nomadic people.

The two major cattle species are *Bos taurus*, the European-type of cattle and *Bos indicus*, the zebu-type humped cattle. The indigenous or local breeds in tropical areas belong to the *Bos indicus* species. They are adapted to tropical conditions and are able to withstand greater heat and a poorer diet than the European breeds. They also have a greater resistance to tropical diseases and parasites.

European cattle (known as exotic breeds), as a result of careful breeding and selection over many generations, can produce high yields of milk and meat. They require good management practices to achieve this in the tropics. In particular their feed must be of high quality, tropical diseases must be carefully controlled, and they must be protected from the extremes of the tropical climate.

There are many advantages to the tropical farmer in cross-breeding his local cattle with European breeds. By doing this the higher yields of European cattle are introduced into his herd without losing the adaptability of the zebu-type cattle.

Tropical breeds

There are several zebu breeds of cattle found in the tropics. Zebu cattle (*Bos indicus*, see Fig. 1.1) have certain characteristics in common:

(a) a prominent hump;

(b) a large surface area per unit of body weight, due to a well-developed dewlap, hump and large ears; this enables them to lose heat quickly;

(c) superior adaptation to high temperatures and intense solar radiation;

1.1 Tropical breeds of cattle

(d) late maturity and rather low fertility;
(e) ability to live on poor quality fodder;
(f) greater stress-tolerance to shortage of water;
(g) resistance to tropical diseases and parasites.

Some of the zebu cattle breeds are discussed below.

Boran

The Boran originates from Somalia and Ethiopia. Although used mainly for the production of beef, it is also a main source of milk in drier areas. The Boran is a fairly large animal with long legs and short horns; it is white, fawn, grey or even red in colour. It has a high ability to utilise rough grazing and to walk long distances to water, without which it can survive for several days.

It is more disease-resistant than the exotic breeds, but not resistant to tick-borne diseases. Through selection, an improved type of Boran has been bred for ranching.

The aggressive maternal instinct of the cows can be a valuable characteristic in protecting their calves in extensive ranching systems.

For dairy purposes the Boran breed should be cross-bred with exotic bulls.

East Africa Zebu

This is the major breed used in East Africa. The animals are small and fine-boned; black, red, fawn or grey in colour, with moderate-sized humps. The cattle are used for milk, meat and draught purposes. They are late maturing, with a low milk production, but are very hardy.

Gir (India)

This breed is mainly used for dairy production, and gives around 1500 kg of milk per lactation. The cattle are red or mottled in colour, usually with a patch of a different colour. The ears are drooping. Males are often used for draught.

Kankrej (India)

This is a very large and powerful draught animal. It has large, sweeping horns, drooping ears and a well-developed hump and dewlap. The colour is grey or blackish-grey with black spots on the legs.

N'dama (Guinea)

A small breed with some resistance to trypanosomiasis, used for meat production. It is usually a fawn colour but may be white or red, and has long lyre-shaped horns.

Sahiwal

The Sahiwal originates from Pakistan. This is a very good tropical dairy breed, although it cannot compete with the exotic breeds from the temperate climates as far as milk yield is concerned. A good cow may give 3000–3500 kg per lactation. The animal is heavily built, short-legged, lethargic in appearance with a reddish or light brownish-red colour, sometimes bearing white markings. The skin is loose and pliable, the hair very fine and glossy. The udder is generally large and often pendulous.

Tharparkar (Pakistan)

A dual purpose breed used for both milk and meat production. Its colour is white or grey and it has small, upturned horns and a fair-sized dewlap and hump.

Australian Milking Zebu

Development of this new breed started in 1950 and it is a cross mainly between Pakistani Sahiwal, Red Sindhi and Jersey breeds. It has been bred in Australia as a tropical milking animal. It is fairly resistant to heat stress and ticks and is a good milker. The breed is expanding slowly in S.E. Asia.

White Fulani (Nigeria)

These are usually kept for beef or for draught. They are usually white with black markings on the ears and tail. The horns are fairly large.

Exotic European breeds

These breeds have been introduced into tropical countries since the turn of the century. The European breeds belong to the *Bos taurus* group (see Fig. 1.2). These breeds are easily recognised because they have no hump. Some are described below.

Ayrshire (Scotland)

A strong and robust animal, characterised by a strongly attached, evenly balanced, well-shaped udder. In colour it is light to deep cherry-red, brown, or a combination of any of these colours with white, or white alone.

The Ayrshire has excellent grazing ability and is predominantly a milk producer. With the rather poor quality of some tropical pastures and the long, dry spells in many areas this ability is very important.

Though the milk quality is better than the Friesian its lower potential yield has resulted in the Ayrshires being gradually replaced by Friesians.

Friesian (Holland)

A strong and heavy animal, with very good dairy and beef characteristics. The colour is typically black and white, all four legs and the lower part of the tail being white. Milk yield and the ability to produce high-quality beef in a short period is excellent.

Guernsey (Isle of Guernsey)

A rather small dairy animal. The colour is a shade of fawn with white markings clearly defined; the milk has a golden-yellow colour. This breed is losing in competition with the other dairy and dual-purpose breeds.

Hereford (England)

A medium-sized, red and white animal. An excellent meat producer, sometimes used in cross-breeding with local cattle.

1.2 European breeds of cattle

Jersey (Isle of Jersey)

A small, fine-skinned, hardy animal. Early maturing and with excellent dairy characteristics. Its colour is various shades of fawn and white; normally has a black ring around the eyes. Bulls can be blackish in colour.

Although the milk yield of this breed is not high, the breed is popular all over the world. There are several reasons for this popularity:

(a) a high percentage of butterfat in the milk compared to other breeds;

(b) the animals are very hardy and adapt well to nearly all kinds of climatic conditions, particularly to high temperatures, low maintenance intake, due to its small size.

Jersey bulls, like many dairy breeds, are unpredictable in their behaviour and care should be taken while handling.

Simmental (Switzerland) and Charolais (France)

The Simmental and the Charolais are very large, big-boned animals, hardy and good meat producers. They are used for cross-breeding.

Table 1.1 Characteristics of Cattle Breeds

Breed	*Approximate weight of mature cow* (kg)	*Use*	*Advantages*	*Disadvantages*
Ayrshire	450	Dairy	High milk yield; meat production fair	Body conformation poor
Boran	400	Beef	Withstands heat and harsh conditions	Late maturing
Charolais	800	Beef	Very good beef producer	Calving difficulties sometimes experienced, especially when used for crossing; high forage requirements
East African Zebu	300	Dual purpose	Very hardy, disease-resistant	Late maturing; low milk production; moderate beef animal
Friesian	600	Dual purpose	High milk yield; good meat production	Relatively low butterfat content; high forage requirement
Gir	380	Dual purpose	Good beef animal	Slow draught animal
Guernsey	410	Dairy	Fairly low maintenance requirement; high butterfat content	Poor meat production; moderate milk production

Breed	*Approximate weight of mature cow* (kg)	*Use*	*Advantages*	*Disadvantages*
Hereford	550	Beef	Good beef producer ideal for cross-breeding with Zebu	Heifers may get too fat
Jersey	350	Dairy	Low maintenance requirement; withstands heat better than other exotic breeds; high butterfat content; small animal	Moderate production of milk; poor meat production
Kankrej	450	Beef	Excellent draught animal	Poor milk producer
N'dama	270	Beef	Good production under poor conditions	Poor milkers and draught animals
Sahiwal	350	Dual purpose	Good meat producer; suitable for semi-dry areas	Relatively low milk yield; at times difficult to milk
Simmental	750	Dual purpose	Good milk producer; good meat producer	Requires high standards of management; high forage requirement
Tharparkar	400	Dual purpose	Good draught animal; very hardy and heat resistant	Milk production useful, but only in conditions of high management
White Fulani	350	Beef	Fattens well with good management	Slow draught animal

Breeding and handling

Artifical insemination

Up-grading the herd

A farmer can increase milk or beef production and profitability considerably by breeding with good bulls. Such bulls are expensive to buy but by using an artifical insemination (AI) service any farmer can upgrade his herd at little cost. Furthermore, by using AI, farmers are able to keep an extra cow instead of the bull they would otherwise need to keep.

Artificial insemination is done by taking semen from the chosen bull (this is done at AI centres) and putting it into a cow when she is on heat. Bulls that are chosen for AI have already sired highly productive daughters.

AI services operate in many countries and particularly in areas of high agricultural potential. Interested farmers should contact their local livestock officer for details.

Milk	*Meat*	*Dual purpose*
Friesian	Boran	Simmenthal
Ayrshire	Charolais	Sahiwal
Guernsey	Hereford	
Jersey		

Breeds available

Farmers should consult their livestock officer about which breed is best for their purposes and farming conditions.

Recognising the heat period

Artificial insemination will only be effective if the farmer or his stockman can recognise when the cow is on heat and ready to be served. (See Fig. 1.3). Also, insemination at the wrong time will be a waste of money. It is not uncommon in the tropics for cows to be inseminated three or four times before they get pregnant, just because the signs of heat are not recognised properly.

The heat period can be very short. More than half the heat periods in a herd will be less than 12 hours long. About half the heat periods are likely to happen at night, increasing the difficulty

1.3 Signs of heat in cattle

of detection. Observation is the key to heat detection. Watch the cows quietly at least three times a day for about 20 minutes each time. The best times to watch the cows are:

(a) early morning, before milking;
(b) early afternoon;
(c) as late at night as possible.

It is useless to watch the cows at milking or feeding times, or when they are being moved, as real signs of heat are unlikely to be seen at those times.

Early warning signs Signs that a cow is about to come on heat are:

(a) licking or sniffing;
(b) chin resting.

Either of the cows may be coming on heat, not just the one that licks, sniffs or rests its chin.

Signs of heat The cow is on heat when:

(a) she stands willingly to be mounted by another cow and does not try to escape;
(b) she mounts another cow from the front; watch her to see if she also stands to be mounted.

This is called the 'standing heat' stage. Additional signs of heat are:

(c) a temporary drop in milk yield;
(d) restlessness;
(e) bellowing;
(f) swollen and reddened vulva lips;
(g) a clear, thin, mucous discharge hanging from the vulva or sticking to the tail.

In tropical countries, the higher temperatures cause more stress to exotic milk breeds of cattle. The cow may come on heat but does not show any sign of it, so the heat period is missed.

Using records If a record of the cow's heat periods is kept, it will be easy to know when to watch her for signs. The cow should not be served for six weeks after calving. After that, each heat period comes every 20–22 days until the animal is pregnant again.

Cattle handing

Calf castration

It is best to castrate unwanted bull calves to ensure that these bulls are not used in breeding. Castration should be done with a knife, an elastrator or a burdizzo. The elastrator, which is the best method for young calves, stretches a strong rubber ring around the scrotum and must be done on the first day of the calf's life. The burdizzo crushes the cords which go to the testes, and is used mainly on older animals.

Training is needed for all these methods, but castration with a knife should only be undertaken by a veterinary officer. There may be an animal health assistant in the district who can advise farmers about animal health and who is able to disbud or castrate animals. Veterinary departments will often train stockmen for larger ranches.

Disbudding

Horns can cause damage and wounds, resulting in stress and reduced milk yield. Horned cattle also require twice as much trough space.

Horns can be removed when the cattle are young by killing the bud with a hot iron at 4 days of age. An experienced stockman should do this. Alternatively, a caustic soda stick can be rubbed on to the horn bud when the calf is less that three weeks old. The operator should wear rubber gloves to protect his hands from the caustic soda.

Drenching

Cattle suffer from infestations of worms: tape, fluke and roundworms. These can be controlled to some extent by rotational grazing, and (for liver fluke) by keeping cattle out of wet grazing areas in swamps and near rivers.

However, to control worms, it is sometimes necessary to treat stock chemically. When an infestation is suspected, a sample of faeces (cow dung) should be sent to a veterinary surgeon for diagnosis. He will then advise the correct medicine. This is normally applied by drenching with a bottle (see under Parasitic diseases, page 59). Care must be taken when drenching to pour the liquid down the throat and not into the lungs, which can result in killing the animal by drowning (see Fig. 1.4).

1.4 Calf drenching. This must be done carefully.

Handling calves

Calves should be handled gently but firmly. Careless handling can cause injury. See Fig. 1.5 for details.

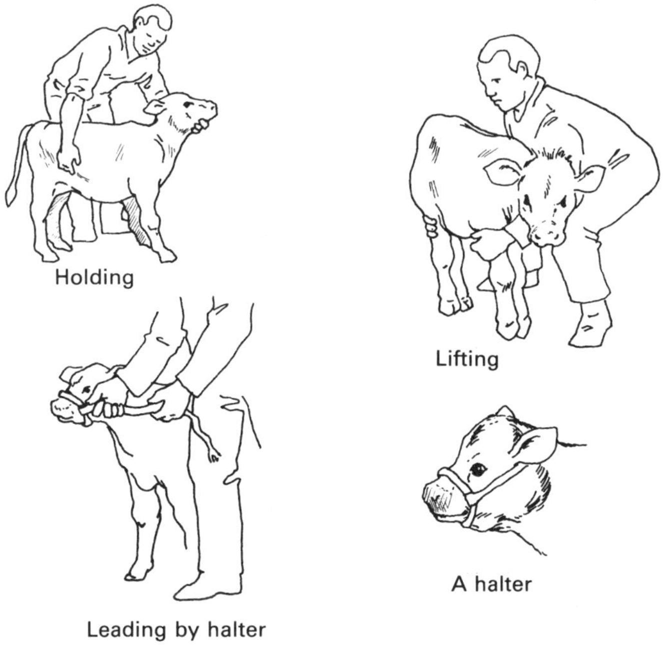

1.5 How to handle calves

Disposal of dead animals

Animals which die on the farm should be buried in deep pits (two metres deep) or burnt. This is important to prevent the spread of diseases, such as anthrax, which are extremely dangerous to animals and humans. Many countries require the Veterinary Department to investigate such deaths.

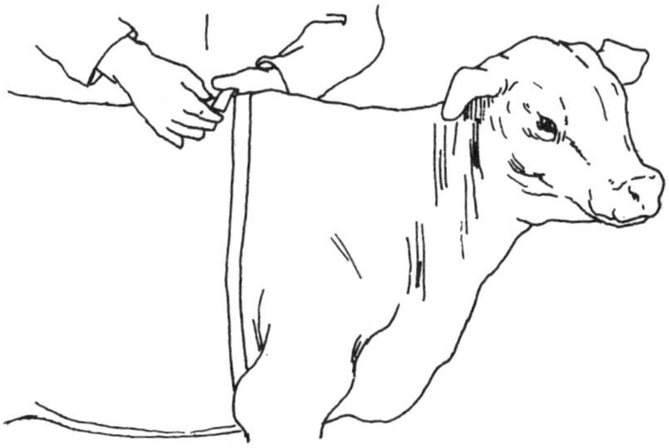

1.6 Measuring the girth of humpless cattle

Estimating the weight of young cattle

Measure the girth, as shown in Fig. 1.6, and read the weight from Table 1.2. This method can only be used for humpless animals.

Table 1.2 Girth and estimated weight of young cattle

Heart girth (cm)	*Weight* (kg)
65.0	26
70.0	32
75.0	40
80.0	48
85.0	56
90.0	66
95.0	77
100.0	89
105.0	103
110.0	116
115.0	132
120.0	150

Estimating the age of cattle

It is sometimes necessary to estimate the age of livestock for which no records are available. The most common method for age determination is by examination of the teeth (see Fig. 1.7). Because the temporary incisors are replaced by permanent incisors at more or less set time intervals, the number of permanent incisors gives a fair estimate of the age of the animal, up to about four years. After four, estimates are usually based on the amount of wear on the permanent incisors. Cattle, sheep and most other ruminants have four pairs of permanent incisors (on the lower jaw only).

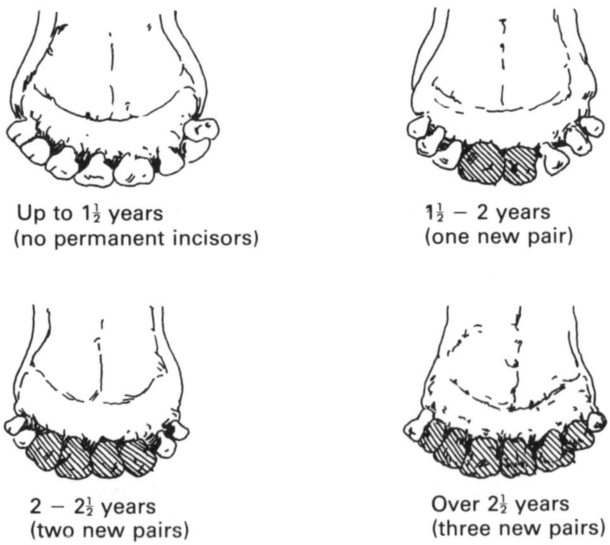

1.7 Estimating the age of cattle by their teeth

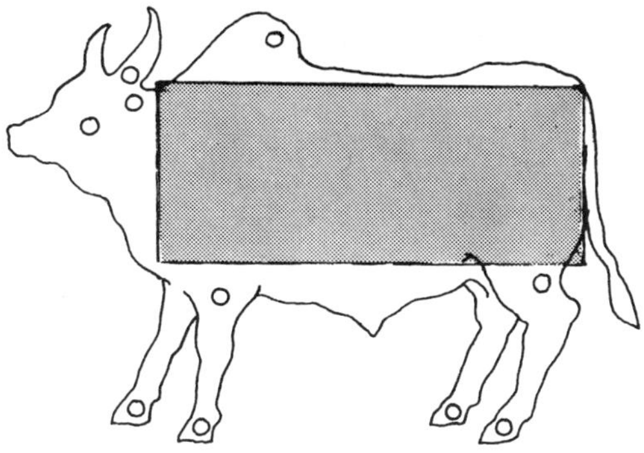

1.8 Branding. Grey area indicates where branding should not be done as it would spoil the hide

Ear-marking and branding

In a large herd it will be necessary to be able to identify individual animals for breeding, marketing, etc. This can be done by clipping the ears, by fastening a metal tag in the ears, or by tattooing a number on each animal. See Fig. 1.8 for suggested branding places. Freeze branding is a good method using liquid nitrogen and a branding iron to bleach hair on the animal.

Hides and skins

Damage to hides

Normally damage to hides or skins occurs during killing and flaying. Some of the most common damage is described below:
(a) bruising: this happens when animals are crowded in the night yard, in crushes, when they are being transported, when the animals are beaten, when they are being branded, and during killing;
(b) damage by warble flies, which spoils the hide by making holes in it;
(c) incomplete bleeding after killing which causes discoloration and damage;
(d) contamination during flaying by manure, blood or soil coming into contact with the hides;
(e) knife cuts which commonly occur when there is a delay in the flaying process; this delay allows the carcass to become cold and flaying then becomes difficult.

It is bad practice to undertake killing and skinning in the evening because the light is poor and the job will not be well done.

Correct skinning

Skinning can best be done with a suitable sharp knife and should be performed immediately after slaughter. Where possible the hide should be pulled off rather than cut off. Contamination should be avoided by removing the stomach and intestines after skinning.

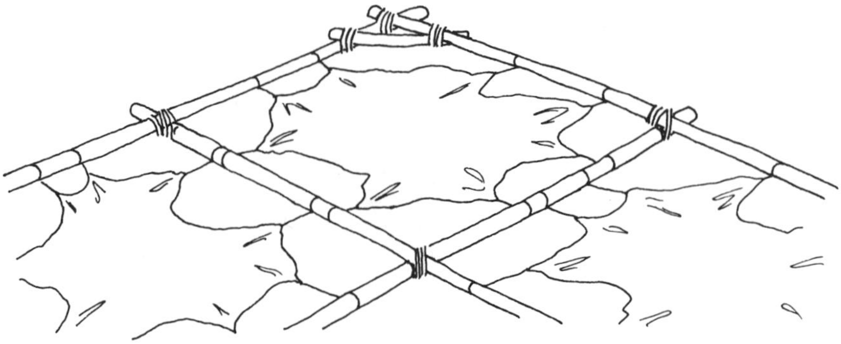

1.9 Suspension drying frame for small skins

Care of the hide

After skinning, the hide or skin should be washed, any remaining flesh removed and the hide neatly trimmed (as described below).

Washing Wash the hide in order to remove blood, manure and other contamination. Proper washing involves spreading the hide or skin on a clean, cemented floor or table. Wash both sides, starting with the hair side first, using clean water all the time.

Fleshing This is the removal of meat and fatty tissues that adhere to the hide or skin. The hide or skin is spread on a clean surface and fleshed by scraping with a sharp, curved knife.

Trimming This is the removal of rough, jagged edges and unwanted parts, e.g. udders and tailbones.

Methods of curing hides and skins

There are two methods of curing hides and skins: air drying and salting.

Air drying Air drying on the ground is simple and cheap but it is not recommended. The main disadvantages of ground drying are that the hides or skins become hard, distorted, light in weight and crack easily. Suspension drying is best (Fig. 1.9).

In suspension drying, hides or skins are fastened on frames and secured by strings. They remain on these frames for 2–5 days, until they are properly dried, then are removed and folded along the back, hair side in. Care must be taken when folding that the hides do not crack down the fold if they are over dry.

Salting Salting is carried out by spreading one hide on top of another, flesh side up, on the floor, Salt is sprinkled on the wet side of each one.

Pest infestation Mice, rats, ants and beetles can cause

considerable damage to stored hides and skins. Use insecticides (e.g. Gamma BHC, Malathion or Carbaryl) to destroy insects. Keep the store clear of rubbish.

Malpractices After drying some people soak the hides or skins and dry them again on the frames to increase their weight and therefore get more money. Sometimes blemishes or marks are covered with clay. These malpractices are easy for the hide purchaser to recognise.

Records

Livestock records

Records are of great value as aids to management and financial control. Stock records are necessary for:
(a) the selection of stock for breeding;
(b) knowing the age for vaccination, weaning, marketing and changes in feed;
(c) knowing when to breed;
(d) identifying animals which produce well or which have chronic health problems;
(e) identifying females that fail to reproduce and males which are not fertile.
Financial records are necessary for the evaluation of the profitability of the enterprise.

Table 1.3 Breeding record

No. Name		Breed	
Purchased from		Date of birth	
Date of purchase	Purchase value	Date of disposal	Disposal value
Sire		Reason for disposal	
Dam			

Table 1.4 Calving record (reverse of breeding card)

Date of service	Bull used	Date of claving	Sex of calf	No. of calf	Weight of calf	Disposal of calf

Table 1.5 Health record (front and reverse)

Vaccinations		Worming	
Date	Vaccine	Date	Remedy

Disease and wound treatment		
Date	Disease or wound	Treatment

If you are running a breeding herd or flock, you will want information on the reproductive history for each dam and sire as well as information regarding the productive ability of the off-spring. On the other hand, if your primary business is fattening cattle you will be more interested in rates of growth and feed consumption. In any type of operation you will want to keep an inventory of livestock showing purchases, sales, births and deaths.

Dairy cattle management

Milk production

Milking qualities

According to average milk yields, the breeds can be listed in order of highest production as follows: Friesian, Ayrshire, Guernsey, Jersey, Sahiwal, Boran.

Milk yields for individual animals vary a great deal even under similar systems of management. However, mature, good-grade cows under good management in tropical conditions will give annual yields approximately as follows:

Friesians	5000 kg
Ayrshires	4000 kg
Jerseys	3500 kg
Sahiwal	3000 kg
Gir	1500 kg
East African Zebu	1000 kg

Some very high-yielding cows, when fed very well, may give double these yields. However, under tropical conditions very high yielders may be under considerable stress and risk to health. Also, much of the extra milk has to be obtained by using expensive concentrates rather than grass or fodder crops.

Breeding age

Cows should be at least 18 months old at first mating, and should have attained two-thirds of their expected mature weight. If they are bred before that it may limit their growth and future production. The gestation period is nine months.

Calving interval

The aim of the farmer should be to produce one calf per cow per year. The cow must therefore be inseminated three months after calving; hence the importance of the farmer's knowledge of heat signs. Failure to recognise the heat means that the farmer has to keep a dry unproductive cow. It also may mean that the cow may calve later at a less suitable time of year. A properly managed cow will calve every year and so be in milk for nine months each year (a period called a lactation). When the milk yield drops, which may happen after 5–8 years, the cow should be sold for slaughter and replaced.

Grazing systems

The grazing systems on farms vary, depending mainly on the type of pasture available, whether these are natural pastures, improved pastures or fodder grasses such as Napier. Natural pastures are cheaper to maintain but have low production in comparison with improved pastures.

The main advantages of improved pastures are increased forage production (often twice as much as natural pasture) and their more favourable response to fertilisers.

With fodder grasses, a higher stocking rate is possible because of the increased quantity of feed available throughout the year. Fodder crops are higher yielding and more drought resistant than natural or improved pastures (see page 35).

Zero-grazing

Under this system the cattle do not graze but are confined in a shed or yard. The system is highly productive, but rather labour-intensive, since the forage must be cut in the fields and brought to the cattle. The cost of the cattle shed is also considerable.

The main feed used on small holdings for zero-grazing is Napier grass and sometimes green maize. By-products such as sweet potato vines, sweet potato tubers, banana stems and garden waste are also fed to cattle. Some 40–60 kg of forage is needed per day per cow. In addition, a mineral lick should be provided and water should always be available.

By feeding on good fodder a dairy cow can produce up to 8 kg of milk per day. For a higher-yielding cow, the fodder must be supplemented with concentrates.

The main advantage of zero-grazing over pasture grazing is that production per land unit is 3–4 times higher. Normally 0.6 ha of good grazing will support one cow, whilst under Napier grass in the zero-grazing system, the same land can support 3–4 cows.

Manure production Zero-grazing can make a major contribution to soil fertility through production of manure. The manure and bedding can be left for some months or removed daily and built into a stack. If the manure is left in the stall there must be adequate bedding added daily to prevent muddy and dirty conditions developing in the stall.

Labour demand In zero-grazing the labour demand is high. One labourer can cut, feed and handle 3–4 cows provided the fodder is grown close to the stall. The labour requirement is for some 600–1000 hours per cow per year, including the growing and harvesting of the fodder.

Calf rearing

Calves are the foundation of the future dairy herd; therefore they should be selected carefully and reared well. Calves are also required to maintain the existing dairy herd numbers, and as an average production span for a dairy cow is five lactations, a 20 per cent replacement of animals will be required each year.

Selection of future breeding cows

Select the best calves from high-yielding cows and registered, proven bulls. A cow can be judged on the milk yield of her first lactation; if it is unsatisfactory she can be culled and not used for further breeding. Where AI is available, it should be used as the cheapest means of breeding, especially for small-scale farmers.

Care of the heifer during pregnancy

For two months before calving, feed the heifer 2 kg per day of concentrates, in addition to her normal grazing. The feeding should be done in the milking shed to help condition the animal to future handling during milking time.

Calving

Calving normally occurs without any difficulty. However when the following signs are noted a veterinarian should be called:
(a) if there is any long delay between the breaking of the water in the uterus and the first sight of part of the young;
(b) if calving takes more than three hours after some part of the young is seen or if the cow is obviously in distress;
(c) if the calf is not emerging in the normal position.

Overdue calving If the date of service is known, the expected date of calving can be calculated. If the calving is overdue by more than one week call a veterinarian. Under no circumstances should anyone except a veterinarian attempt to assist the birth by pulling out the young, as both calf and dam may be seriously injured or even killed by an inexperienced person.

Bull calves Bull calves of the Jersey and Guernsey breeds are not economical meat producers, so they should be slaughtered early, unless they are required for breeding. The usual time is after the flow of colostrum from the mother has ceased, when the calf is about one week old. Friesian and Ayrshire bull calves can be reared for beef if there is sufficient fodder available.

Colostrum Ensure that the calf gets the colostrum that the dam produces in the first few days of lactation, as it is rich in essential vitamins and minerals. The mother's immunity to some diseases is transferred to the calf through the colostrum, but only during the first 36 hours of the calf's life. Therefore it is essential to feed the colostrum at this time. It also has a laxative effect and so helps the calf get rid of its first faeces.

Housing For the first two months the calf should be kept in an individual stall measuring about 2 m × 1.5 m. In warmer areas a

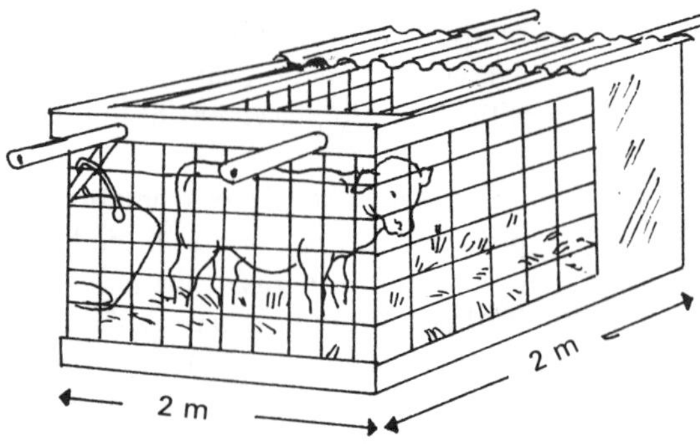

1.10 Movable calf pen

movable calf pen provides adequate shelter for the young calf (Fig. 1.10). In colder areas, the pen should be permanent, warm and free from draughts, and should be kept clean. It should have a slightly sloping floor so that liquids can drain off. After the first two months, it is a good idea to have movable pens placed out on clean pasture to help prevent worm infestations.

Feeding Leave the newly-born calf with its dam for the first three days to suckle. After this, place the calf in a clean, disinfected pen with dry bedding and for the next four days bucket-feed with milk three times a day. Feed the milk at body temperature, immediately after milking. After seven days provide clean drinking water and hay or young, fresh fodder, in addition to the milk.

At eight weeks put the calf to pasture during fine weather. Calves should graze in fresh paddocks ahead of the main herd, because of the danger of worm infection. The paddock should have been free from grazing for at least three weeks; the sun kills worm eggs. Alternatively, calves can be tethered and moved to a fresh place twice a day.

When the calves are put to grazing, provide a mineral lick. If possible, provide newly-born calves with doses of vitamins A, B, C and D (following the tables and instructions on the packet), or give 300 ml of cod liver oil.

Scouring Scouring is the passing of watery unhealthy-looking faeces. Calves should be watched very carefully for scours as it may be fatal. Treat scouring as follows:

1st day: no milk at all, just provide clean, boiled water three times a day.
2nd day: half the normal milk ration plus water and glucose.
3rd day: half the normal milk ration plus water and glucose.
4th day: return to normal ration.

Rearing systems

Farmers have the choice of three rearing systems for calves.

Natural The calf suckles from the dam until it is weaned at four months. The cow is only milked if the calf does not take all the milk. This is a useful system for beef breeds and low milk producers.

Substitute mother One gentle, old cow can mother up to four calves by controlled suckling (thus releasing their mothers for milking); calves suckle quickly so a calf should not be left on the cow too long or it will finish all the milk.

Bucket The calf can be fed from a bucket (see Fig. 1.11) thus leaving most of the milk for human consumption. This is very suitable for cows giving a good yield.

1.11 Teaching a calf to drink

Weaning

After the colostrum period the farmer has a choice of two options.

Early weaning at 10 weeks About 225 kg of milk is consumed during this period, plus 60 kg of concentrate and 40 kg of maize meal. The supply of concentrates starts in the sixth week and ends in about the sixteenth week.

Late weaning at 16 weeks The milk consumption during this period is about 390 kg. No concentrates are supplied.

Starting on concentrates Young calves often do not recognise concentrates as something edible, but can be encouraged if a handful of concentrates is sprinkled on the milk in the bucket. When the calf empties the bucket and is consuming the last of the milk it will also eat the concentrates.

Availability of roughage Good hay should be available at all times to provide bulk in the diet and to promote rumination.

Table 1.6 Bucket-feeding schedule (kg/day)

Age in weeks	*Late weaning*	*Early weaning*		
	whole milk	*whole milk*	*concentrates*	*maize*
1	colostrum	colostrum	—	—
2–3	3	3	—	—
4–5	4	4	—	—
6–7	5	5	0.25	—
8–9	4	4	0.50	0.25
10–11	3	—	0.75	0.50
12–13	2	—	1	0.75
14–15	2	—	1	1
16		—	1	1

Note: A commercial milk replacer mixed with warm water can be used after the first two weeks; it is usually cheaper than whole milk.

Changing feeds Any changeover from one type of feeding to another should be done gradually, preferably over a period of at least a week. Sudden changes will lead to digestive disturbances which can cause scouring and lead to other, more serious conditions and sometimes death.

Calf sucking After calves have been fed, they will tend to continue sucking anything they get in their mouths including the ears, navels and udders of other calves. This is harmful to the sucking calf as it sucks in air and becomes bloated, and may also cause permanent damage to the other calf, particularly to the udder. If the calves are given dry food such as grain after their milk feed, sucking will be greatly reduced. Some calves will continue to suck regardless and should be isolated from the other calves.

Dairy cattle housing

Dairy shed

Dairy sheds are essential in wet areas so that cows can be milked in some degree of comfort and cleanliness. Permanent dairies are normally built on bigger farms, whereas on small farms a movable dairy shed has proved to be cheap and useful. The permanent dairy shed should be cleaned daily after milking and the movable shed should be moved to another place every 2–3 days.

Cattle crush

Each farmer should have a crush on his farm unless there is a cooperative crush close by. This is necessary in order to control animals securely when they are being sprayed or for any kind of examination or treatment by a veterinarian (see Fig. 1.12).

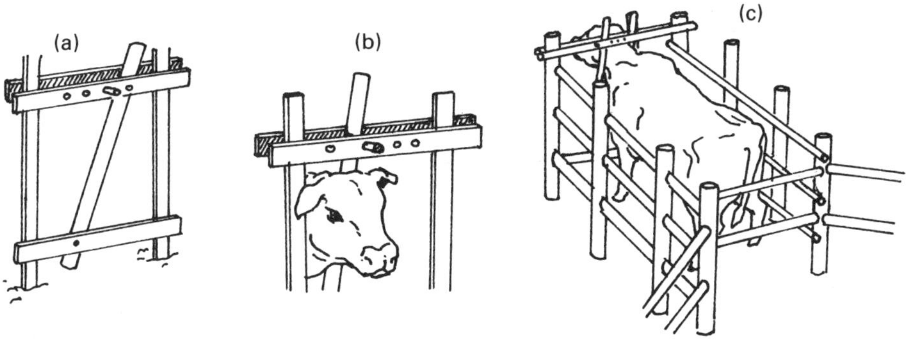

1.12 Controlling cattle:
(a) and (b) head yoke, (c) holding crush

Zero-grazing unit

Such a unit can have a corrugated iron or thatched roof and may have an outside yard, although this is not necessary (see Fig. 1.13). The unit can be cleaned out each day. However, if there is sufficient bedding available then a deep litter system can be followed. In this system the unit is cleaned out only once or twice a year. More straw is added every few days so that animals always have dry bedding.

Dairy cattle feeding (concentrates)

Concentrates are high-protein/carbohydrate/mineral mixtures which are produced commercially; they can also be mixed by the farmer himself.

Grass feed Since concentrate feeds are expensive most of the nutrients required by cows should be produced on the farm as grass, fodder crops, silage, etc.

Feeding of concentrates A production of 10 kg of milk per day per cow can be achieved from good pastures. However, cows producing more than this require extra food in the form of concentrates. If the pasture is poor then additional concentrates will be needed to maintain milk production.

Mixing concentrates Whether to use home-made concentrates depends mainly on the availability of the components. In general it is probably more satisfactory to purchase a commercial concentrate, unless the price is too high. The ingredients for a home-made mixture could be as follow:

60 kg cereal (maize, barley and/or wheat, crushed or milled)
25 kg oilseed cake (cotton, groundnut, sunflower, coconut)
15 kg milling by-products (bran, maize germ-meal)
2 kg mineral mixture
102 kg total

Plan

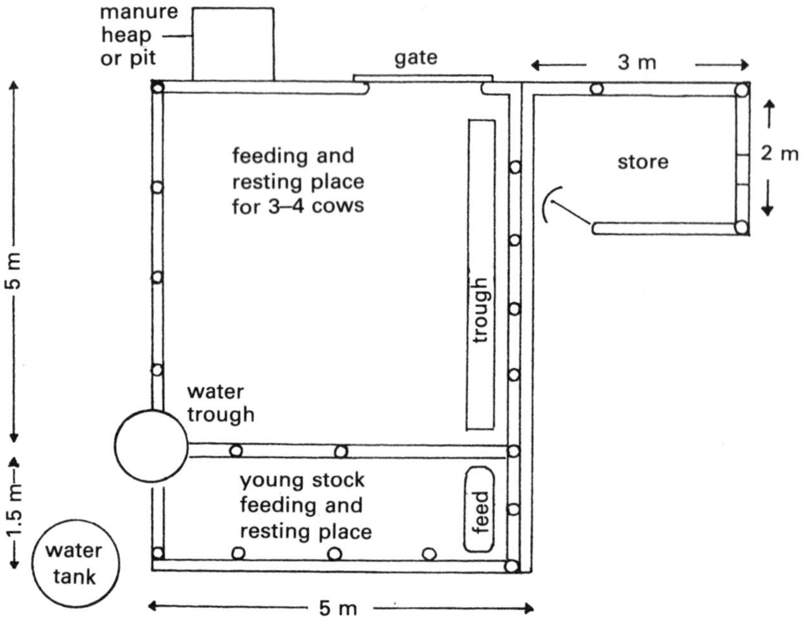

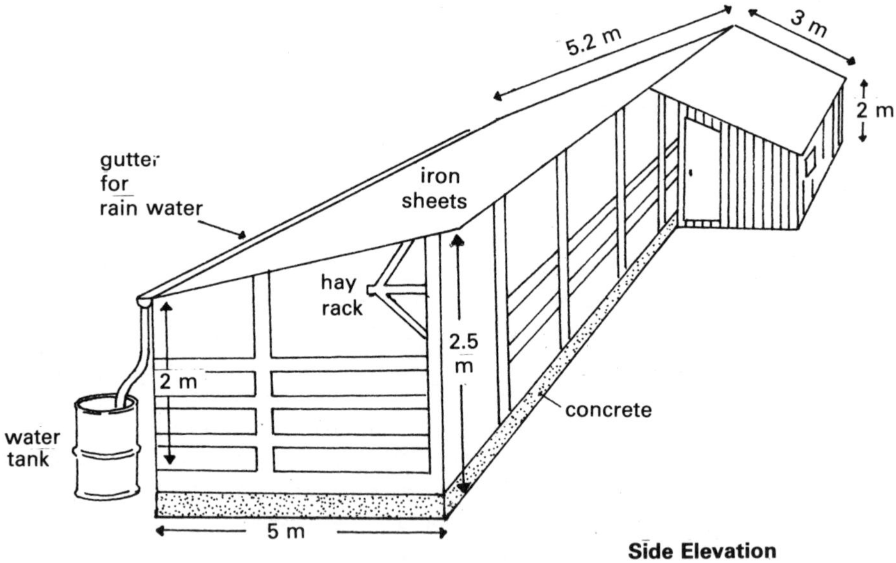

Side Elevation

1.13 Zero grazing or stall feeding unit. The length of the unit will depend on the number of animals to be housed.

Dry heifers and cows (in calf) Dry cows should be fed well to keep them in good condition. However, they should not be allowed to become too fat since this causes difficulties at calving time. A dry cow requires no concentrates if she is on good pasture, but should be fed 1–2 kg per day on poor pastures.

Steaming up before calving During the last two months before calving heifers and cows should receive about 2 kg of extra concentrates per day. This process is called 'steaming up'. During these last months the foetus is growing fast, and also the heifer or cow has to build up some extra energy to be able to withstand the requirements of the coming lactation.

After calving (body maintenance) As the young heifer still has to grow, she needs extra feed. First-calvers should receive 2 kg of concentrates and second-calvers 1 kg per day, in addition to other concentrates fed for milk production.

Cows in milk Cows in milk should be grazed on good pastures. If grazing is short during the dry season, they should be fed with silage, hay or other fodder crops, or the milk yield will drop.

Table 1.7 Efficiency of pasture for milk production

Pasture condition	*Pasture is sufficient to produce* (kg milk per day)
Poor	3
Medium	6
Good	10
Very good	13

Feed 1 kg concentrates for each additional 3 kg of milk produced over the figures above.

After the cow calves the amount of milk produced per day will increase until it reaches a peak about six weeks into the lactation. During this period an extra 1 kg of concentrates should be supplied daily to enable her to build up her production. Three months before calving, dry off the cow by ceasing to feed her concentrates and feeding more hay. Only milk her once a day for a few days then once every other day. The cow will then dry off.

Minerals Minerals are a very important component of cattle feed. Mineral deficiencies result in deformation of the bones, poor growth and low fertility. Minerals are essential for lactating animals because of the high drain of minerals from the cow's body during milk production. It is advisable to provide cattle with mineral bricks or add a mineral mixture to the concentrate. These mineral mixtures contain common salt, calcium, phosphorus, magnesium, copper, cobalt, iron, potassium, iodine, zinc, manganese and sulphur. Common salt alone is insufficient. For dairy cows, a minimum of 50 g per day of mineral mixture should be given, or a mineral lick can be available at all times.

Water supply A good water supply is vital to successful cattle husbandry. Dairy cows should be given water three or four times daily. The daily water requirement for a dairy cow is about 35 litres, plus three litres per litre of milk produced.

Farmers who have access to rivers or swamps should bear in mind that such a source is, in most cases, infected by disease organisms such as flukes and worms, which are a danger to the health of the animals.

The drinking place should not be too dirty or muddy, so a dry access to the watering place should be constructed. This can be done by making a stone or concrete area where cattle can stand to drink, and by preventing them from entering the water by building a small wall or fence. A good method is to bring some water from a stream by means of a furrow or pump, and channel it into a cattle trough.

Some farmers may obtain water from wells. The top of the well shaft should be 1 m above the ground and have a concrete surround and a wooden cover to keep out dirt and for the safety of young children.

Corrugated roofs on houses or stores can be used as rain catchments and connected with water tanks. To establish this type of water supply requires a relatively high investment. Such a catchment will be most useful but unless the roof is very large it will not provide all the water requirements for the year.

Milking

The aim of having a high standard of hygiene in milk production is to supply milk in clean, fresh, wholesome condition, free from objectionable odours and flavours, and to prevent the spread of diseases.

Clean milking system

Under warm conditions, the number of bacteria present in milk can increase very rapidly. A single bacterium can multiply itself up to a million in a few hours, and the milk can become unusable in a short period, especially in hot and humid conditions. The following are the main hygiene requirements to ensure clean milk production.

(a) Milking shed. The milking shed should be cleaned after each milking.

(b) Human infectious diseases. People suffering from infections or diarrhoea should not milk the cows.
(c) Washing hands. Before starting milking, the milker should wash his hands and arms with soap and hot water, or disinfectant.
(d) Wash udder. Before milking, any mud or dung should be washed off the udder with water. Then the udder should be washed using a clean cloth and clean water which has disinfectant added to it. Use two clothes alternately for washing the udders. Leave one in the disinfectant whilst the other is in use.
(e) Reducing bacterial content. The first draw of milk from each teat should be thrown away as it has a very high bacterial count.
(f) Mastitis. Cows with mastitis should be milked last to prevent the spread of the infection to other cows, as mastitis can cause a large reduction in milk yield. Milk from sick cows, including those with mastitis, should not be sold or fed to other stock. (See page 58.)

It is emphasised that the best way to prevent mastitis is by its prevention and strict attention to hygiene when milking. The first milk hand-drawn from a cow should be studied for tell-tale traces of blood clots as an early warning of an attack. A small cup can be used for this (a strip cup). The use of badly regulated milking machinery is an obvious source of injury leading to infection.
(g) Teat dipping. After milking, use a teat dip containing a suitable antiseptic. This helps to prevent the spread of mastitis.
(h) Filtering and cooling. After milking, the milk should be removed to a separate room, filtered and stored in a cool place.
(i) Washing utensils. All utensils which have been used for milk or milking must be washed and scrubbed with hot water and detergent and then rinsed with clean water, before being put on a rack to dry, preferably in the sun.

Otaining good yields

With high-yielding cows it is essential to practise good milking techniques. The following are some rules to be applied in milking:
(a) milk at regular intervals, at the same time morning and evening;
(b) maintain peace and quiet during milking time;
(c) milk quickly and evenly;
(d) empty the udder thoroughly at each milking.
A conditioned reflex in the cow relaxes certain muscles and permits milk 'let-down'. Thus a regular routine should be followed as, for example, the rattle of the milk bucket, supplying feed, waiting in the yard or washing the udder may all set off the conditioned reflex.

The milk let-down lasts only 5–7 minutes, so quick, efficient milking is required. The cows should not be upset just before or during milking as this will prevent let-down.

Do not strip with the thumb and forefinger but use the following technique: grasp the teat firmly at its base between the thumb and the forefinger and close the other three fingers in turn from the top so that the milk is squeezed from the teat into the bucket. Do not pull on the teat, as this can cause mastitis.

The milk yield of a cow reaches its maximum 6 weeks after the birth of a calf and then gradually decreases over the next 8–10 months.

Milk records

If each cow's milk is weighed, either every day or one day each week, then production can be recorded. This can give the cow's lactation, which is the total milk she gives from the birth of the calf until she dries off.

Milk records allow the farmer to see if a cow's milk yield is satisfactory. If it goes down suddenly, the cow may be sick, coming on heat, or contracting mastitis.

Cows with poor milk production over the whole lactation can be culled (sold off) and be replaced with new stock from a reliable source.

Measuring milk To keep records, the milk has to be measured. It may be weighed in kilogrammes using scales, or measured in a special bucket marked in litres.

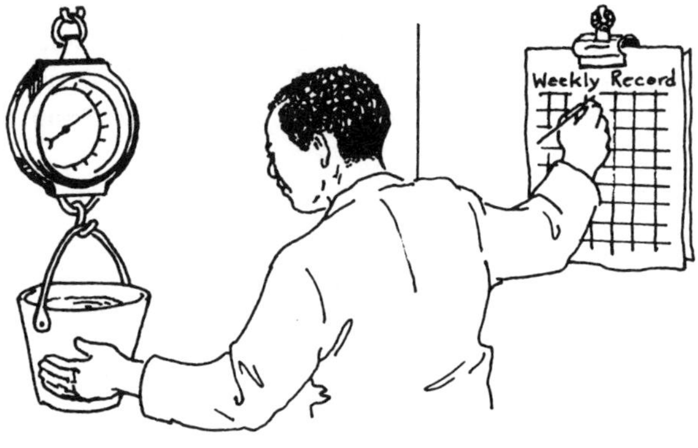

1.14 Recording milk yields

Grazing, fodder crops and fencing

Grazing methods

There are three different methods of grazing: continuous grazing, rotational grazing and strip grazing.

Continuous grazing The cows are always kept in the same field. This is a poor method, as the grass is continually being walked on, and also the worm infestation on the pasture will build up.

Table 1.8 Milk recording chart

Weekly milk record (kg) *Month* .... *Year*					
Names of cows					
Week	Morning or evening milking				
	AM				
	PM				
	AM				
	PM				
	AM				
	PM				
	AM				
	PM				
	AM				
	PM				
	TOTALS				

Rotational grazing This is a better method of grazing. The farm should have a certain number of fields (by preference, five or more). The cows are moved from one field to the next in rotation, which allows the fields time to recover.

The resting time after cows have been grazing a field should be at least four weeks. Grazing too early inhibits growth and root development of the grass plants, although grasses such as Star grass can withstand heavy grazing. Undergrazing, however, allows plants to become too mature and coarse.

The maximum grazing time for each field is about one week, in order to avoid worm infection. Ten cows need approximately one hectare of good new grazing per week. On many farms it is not feasible to have five fields and in order to have good grazing, such farmers should practice strip grazing.

1.15 Strip grazing. Allow cattle to graze until the grass is eaten, then move the fences to a new area.

Strip grazing This is an intensive way of rotational grazing. The animals are confined to a small area sufficient for a few days' grazing, and the fence is then moved (see Fig. 1.15). Sometimes an electric fence is used. The results are more uniform grazing and consequently little wastage of grass. Strip grazing requires more labour and more watering points, but experiments have shown an increase of 15–20 per cent in efficiency of pasture use.

Division of cattle for grazing

On large farms it is important to divide the dairy herd into groups according to age and production. For smallholders this is not feasible because of the small number of animals.

Graze each pasture first with the calves up to one year of age, then follow with the milking herd and heifers in calf (two months before calving) so that they also receive good grazing. Finish up the grass with dry cows and heifers over one year of age. This system will minimise the risk of worm infection for young calves and provide them and the milking herd with the best grazing.

Treatment of pastures after grazing

The smallholder can either tether or zero-graze his calves so that the risk of worm infection is minimised.

After the cattle leave the pastures the farmer should determine whether the grass has been uniformly eaten. If not, the field should be mowed to allow an even re-growth. The best time to mow is just before each rainy season.

An application of 100 kg/ha of a nitrogen fertiliser should be applied early in the rainy season. This will stimulate grass growth. Cattle droppings can be broken up and spread around the pasture with a hoe or harrow, otherwise they will cause coarse clumps of grass to grow. This should be done as soon as the cattle have been moved from the field. This also exposes parasite eggs to direct sunlight and can thus be a useful control method for parasites.

Grass ley management (temporary grasslands)

For good grass leys the soil needs to be well-drained. The rainfall should be more than 800 mm/year and well distributed over the year. The seed bed should be as level as possible for easier mowing of the pasture later on.

Grass seed is small so it is necessary to prepare the seed bed to a fine tilth. A grass ley is likely to be left down for 3–6 years.

Considerable research has been undertaken on legumes and grass/legume mixtures in recent years. The results have shown that on well-managed mixed pastures, considerably higher rates of weight gain or milk production have been achieved. Useful combinations are Centro with Guinea, Elephant, Star or Pangola grass; and *Stylosanthes guyanensis* with Guinea grass. Seek advice from research stations or from companies selling grass and legume seed.

Planting or sowing

Pastures can be established either by using seed, rootsplits or rhizomes, depending on the variety of grass to be planted. Young grass is susceptible to drought, and so pastures should either be irrigated or sown when rain can be expected. There are three methods of sowing described below.

Undersowing Grass can be sown under a cereal crop such as maize or wheat. The sowing of the grass takes place after the weeding of the cover crop. It is important that broadcast grass seed rates are increased by 50 per cent compared to direct sowing rates and that the cover crop is removed from the field as early as possible.

Direct sowing This is done by seed drill or by hand. Because the volume of seed is small, it is normally sown mixed with fertiliser or with sand. The seed rate varies with the grass species. The recommendation is normally 5–10 kg/ha. When sowing grass seed it should not be buried deeply but only dropped on the surface of the soil. After sowing it is useful first to roll the land, then use a very light harrow or drag a thorn bush over the ground in order that the seed is very lightly covered.

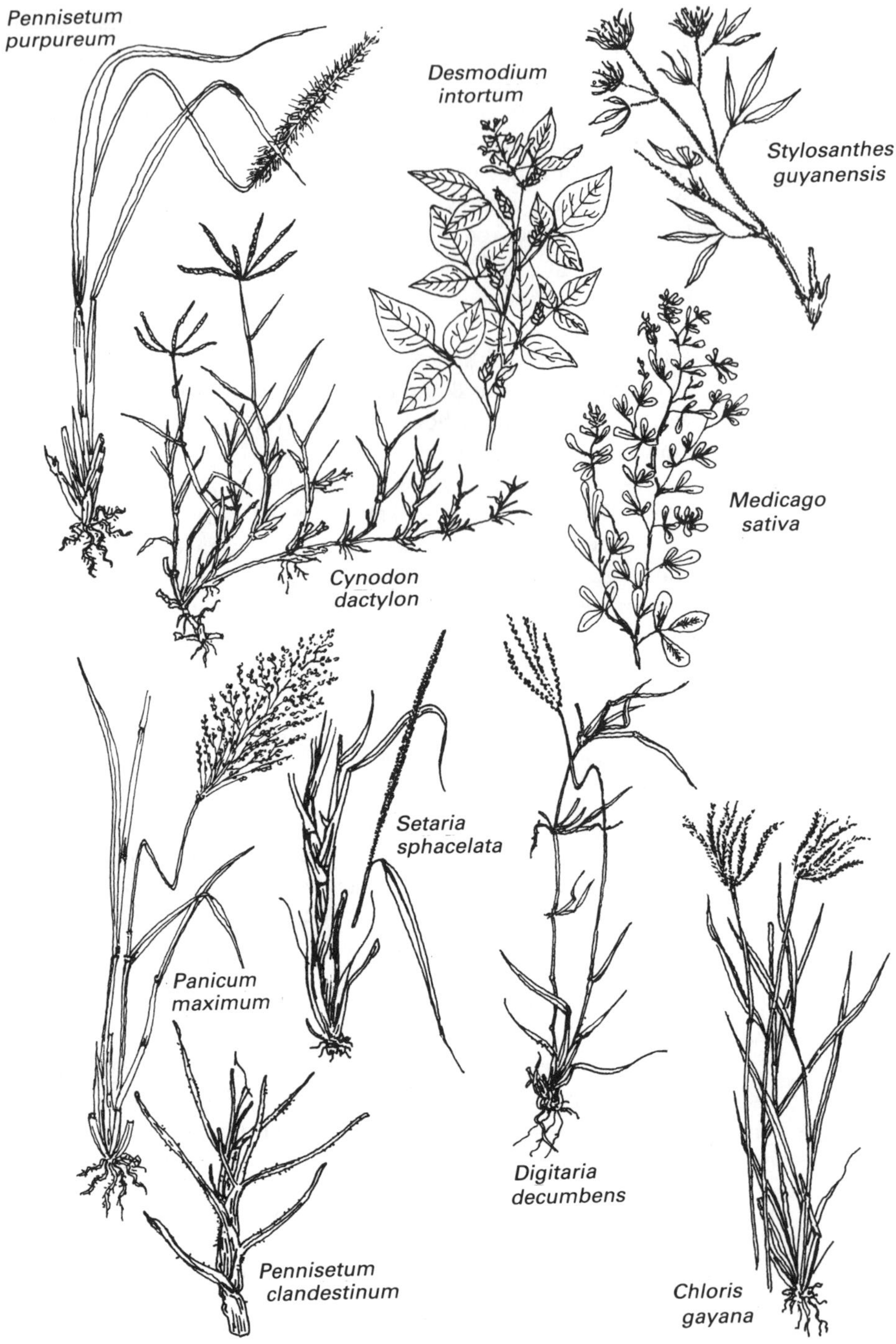

1.16 Grasses and legumes

Oversowing Oversowing is the introduction of improved pasture species into unimproved grassland. It is commonly practiced in steep, unploughable, high-potential areas, by broadcasting seed or planting cuttings in the existing pasture. The productivity of the grassland can be improved at low cost.

Grazing management

In the first year grazing should be light. Freshly established leys should not be grazed under wet conditions or they will be spoilt. The ideal height of the grass before grazing is around 30 cm.

Grass is highest in bulk just before flowering; it is also more palatable than after flowering and has a higher protein value. If pastures are under-grazed there is the danger of the grasses becoming coarse and too mature. Overgrazing, on the other hand, will result in damage to the grass plants and low production, with the grasses being killed or very slow to recover. Useful grasses and legumes are illustrated in Fig. 1.16.

Table 1.9 Common grasses and legumes

Botanical name	*Common name*	*Environment*	*Use and propagation*
Pasture grasses			
Cynodon dactylon (creeping perennial)	Star grass Bermuda Bahama Couch	High and medium potential areas	Tough, robust grass withstands trampling; good pasture grass; difficult to eradicate; plant by cuttings
Panicum maximum (tufted perennial)	Guinea grass	Warm, humid climate; can withstand dryish conditions of rainfall of around 870 mm	Very useful pasture grass; seed required 4–5 kg/ha
Cenchrus ciliaris (tufted perennial)	African foxtail	Hot, tropical and subtropical areas	As pasture or for hay; seed required 4 kg/ha, but first keep seed for 2 years
Chloris gayana (perennial)	Rhodes grass	Medium dry areas up to 2700 m altitude	Pasture grass; seed required 5 kg/ha
Digitaria decumbens (perennial, runner)	Pangola grass	Wet, coastal areas	Pasture, high-yielding; plant with cuttings
Hyparrhenia rufa (tufted, tall perennial)	Hyparrhenia	Unimproved grasslands	Tough, drought-resistant, natural pasture
Paspalum notatum (perennial, rhizome)	Bahia grass	Fairly drought-resistant	Useful for conservation or waterways; plant with cuttings

Botanical name	*Common name*	*Environment*	*Use and propagation*
Pennisetum clandestinum (perennial)	Kikuyu grass	Needs at least 1000 mm rain and grows well at 2000 m altitude and over	Pasture grass; plant with cuttings
Setaria sphacelata (tufted perennial)	Nandi setaria	Fairly tolerant of waterlogging; does best in high rainfall areas at altitudes of 1300–3000 m	Pasture grass; seed required 5 kg/ha, or by cuttings
Fodder grasses			
Pennisetum Purpureum	Napier grass	Warm conditions and with rainfall more than 1000 mm; can withstand dry spells; high-yielding and requires good soil and fertiliser	Use for green chop or silage; can be grazed with care; plant with cuttings
Tripsacum laxum (tall perennial)	Guatemala grass	Moist and swampy soil, fairly drought-resistant	Green chop and silage; plant with cuttings
Centrosema pubescens (creeping, climbing)	Centro	Wet areas but will withstand a long dry season	In a pasture together with tall grasses; seed (innoculate)
Desmodium intortum (trailing, deep-rooted)	Green-leafed Desmodium	Medium potential areas	In a pasture; seed (innoculate)
Desmodium uncinatum	Silver-leafed Desmodium	Medium potential areas	In a pasture; seed (innoculate)
Glycine wighti (trailing perennial)	Glycine	Medium potential areas	In a pasture; seed
Stylosanthes guyanensis (deep rooted perennial shrub)	Stylo	Warm climate, drought-resistant; will grow in low fertility conditions	In a pasture; seed
Trifolium spp (perennial, runner)	Clover	Grows over 1700 m with good rainfall	In a pasture with low-growing grasses; seed
Medicago sativa	Lucerne	Subtropical, and will withstand some drought; good under irrigation conditions	High yielding; best for haymaking; 4–5 cuts can be taken each year; needs heavy fertilising; seed (innoculate)

Fodder crops

Fodder is the term used to refer to crops used for feeding animals. It can be fed fresh as green fodder or conserved as in hay or silage. By conserving crops when they are in abundance and feeding them when pasture is not available, the effective stocking rate of a farm can be increased.

Types of fodder are hay, maize silage, Napier grass, sorghum and sweet potatoes. Below are some brief notes on the production of different types of fodder. A farmer should obtain expert advice before attempting to produce hay or silage. For the small farmer, other fodder crops are usually preferable.

Hay

Good hay is excellent feed for dairy cows and calves, and is made from young, green grass cut before the heads mature and then dried quickly to retain its food value. Too much rain at haymaking time will result in poor quality hay.

Cutting Grass is usually ready for cutting in the middle of the rains. This makes it difficult to dry and it is often spoilt. However, if the grass is lightly grazed early in the rains then the time of cutting will be delayed until the end of the rains, thus giving a much better chance to make good hay.

Grass cutting requires considerable labour. For large quantities, a tractor or ox mower is necessary.

1.17 Methods of stacking and storing hay

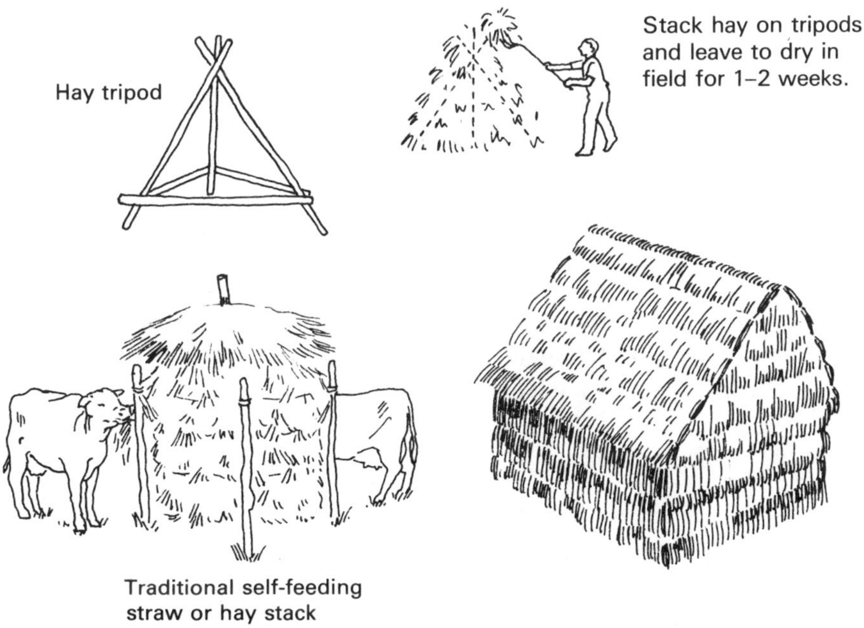

Drying The cut grass should be left on the ground for three or four days and turned several times to speed drying. Sometimes the grass is then built into small heaps to finish drying if there is a danger of rain. Once the grass is dry it can be made into a 'stack', which should be well thatched, or it can be stored in a shed.

The hay should be dried with as little turning as possible to prevent leaf loss and it should not be left in the sun longer than necessary. If a handful is twisted and no moisture is seen then it is ready to store.

Maize silage

Maize silage is used for feeding dairy cattle and on beef feed-lots, in large-scale farms. Advice on the most suitable variety for silage should be sought locally. It is difficult to make silage without mechanisation because of the weight of the crop, so its use is restricted to the larger producer.

Manure and fertiliser Manure may be ploughed in before planting the maize seed. Double superphosphate is applied during planting at the rate of 100 kg/ha. Top dressing with nitrogen when the plants are 25 cm tall at the rate of 50 kg/ha is recommended.

Spacing The spacing will vary slightly for different varieties, but will be approximately 75 cm × 25 cm. About 35 kg/ha of seed will be required.

Harvesting Harvest when the grain is at the firm dough stage. Large-scale farmers use machinery such as forage harvesters and trailers for cutting and carting. Silage is made by compressing the chopped fodder in a silo. If the air is excluded then the fodder will be preserved and can be used months later when there is no green feed available.

Silage There are three types of silo:
(a) heap or stack silo, made on top of the ground on a well-drained site; there is likely to be considerable wastage around the sides;
(b) pit silo from which it is more difficult to remove the silage for feeding, being low down;
(c) bunker silo which may be either dug into a slope or have walls built above ground; it is expensive but the best type. (See Fig. 1.18.)

Air control in the silage Air should be prevented from entering the silo as oxygen will cause oxidation of sugars in the forage, and the silage will be of poor quality. If too much air is present the silage becomes too hot and the overheating causes the silage to

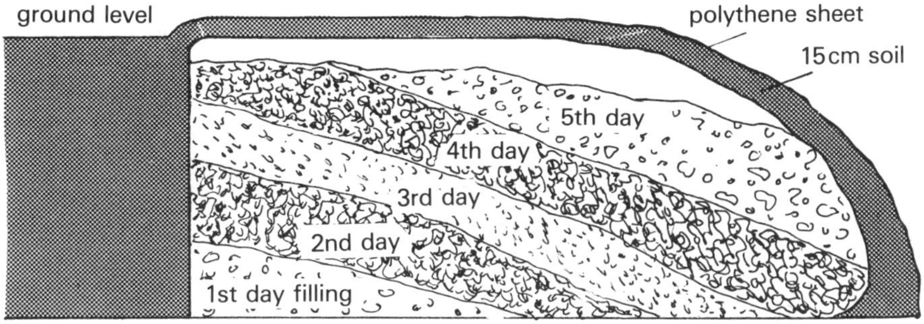

Bunker-type silo

Silage must be consolidated between loads with a tractor or cattle.

1.18 Types of silo

become dark brown. Good silage is sweet smelling and green to light brown in colour. The amount of oxygen trapped inside the forage is minimised by:

(a) chopping the forage as finely as possible;

(b) consolidating the forage well when the silo is being filled, by trampling by foot, with oxen, or by frequently running a tractor over it;

(c) preventing the entry of oxygen by covering the silo with a plastic sheet and sealing the edges with soil placed over the plastic.

Protection from rainwater When the silo has been filled, the last few loads of forage should be used to form a ridge in the centre which will assist water drainage after the heavy gauge polythene cover has been put in place. A depth of 15 cm of soil should be put on top of the cover to weigh it down and help keep out air.

Use of molasses Maize silage does not require the addition of molasses. Forage low in sugars, e.g. grass or sorghum cut at early stages of growth, should have molasses added during ensiling.

Forage with a high moisture content should be wilted for half a day before ensiling and also have molasses added.

Molasses should be mixed with water at the ration of one part of molasses to two parts of water to make it easier to apply. A watering can with a sprinkler attached may be used; 10 litres of this mixture should be used for each cubic metre of grass.

Napier Grass (Elephant grass)

A high-yielding fodder crop is Napier grass and it is widely grown.

Yields Napier grass can reach a height of 4 metres under favourable conditions, and when cut or grazed regularly, gives yields of green fodder of 75–150 tonnes per hectare. In areas with dry spells not longer than 3–4 months it remains green throughout the dry period.

It can be planted on large farms as a field crop or on small farms along fence lines, in odd corners or in other suitable places.

Climatic range It grows well at altitudes up to 2500 m and will grow more slowly at higher altitudes. Rainfall should be over 900 mm per year for good growth.

Varieties Your local research station should be able to provide you with information about the most suitable varieties available.

Soil preparation Napier responds well to soils of good fertility. Soil preparation should be as for maize. Couch grass should be eliminated by hoeing and harrowing. The roots should be removed from the field.

Planting Stem cuttings or rootsplits are used as planting material; the latter gives quick establishment. Cuttings with at least three nodes should be planted at an angle, with two-thirds in the ground. The spacing should be 1.5 m between the rows and 20 cm between plants in the rows.

Fertiliser At planting time 150 kg/ha single superphosphate should be applied. A top dressing of 200 kg/ha sulphate of ammonia can be applied in three applications during the year.

Cutting The best time to cut Napier grass is when it has reached a height of about one metre. A stubble about 25 cm high should be left. Competition from weeds, especially couch grass, has to be avoided by weeding regularly, at least twice a year.

Grazing management Napier grass can be grazed and the most productive period is when it is 3–4 years old, after which the yields decline considerably. The first grazing usually starts 3–4 months after direct planting and should be light. Later grazing should be done at a height of 50–70 cm; this is before the stems have become too coarse. However, Napier grass cannot withstand continuous grazing so it should be given time to recover and to reach the above-mentioned height.

Diseases Napier grass is little affected by diseases. Occasionally leaf rust occurs which can be successfully discouraged by heavy grazing.

Sorghum

Varieties include Haygrazer, Chowmaker (hybrids), Sudan grass, Velvet sorghum, Columbus grass and Bulrush. Spacing is at 15 cm × 60 cm. Sorghum will grow in sandy soils in dryer areas.

Fertiliser Apply 100 kg/ha double superphosphate in the rows before planting and top dress with 100 kg/ha of sulphate of ammonia after cutting.

Harvest Fodder sorghums can be cut every 6–8 weeks, making five or six cuttings. The plants should be cut only after attaining a height of 60–70 cm (when younger they have high levels of prussic acid which is poisonous to cattle).

Pests and diseases Leaf blight and shoot fly attack sorghum but are not serious problems and spraying is not ncessary. Head smut attacks the heads if they become over-mature. Control by cutting or grazing at an earlier stage.

Sweet potato

Grow on fertile soils and in areas which are not affected by frost.

Planting Vine cuttings and tubers are used for planting. Varieties include high vine- and high tuber-yielding types.

Spacing Spacing is 90 cm × 30 cm for tuber production and 60 cm × 30 cm for vine production.

Manure Manure can be applied during land preparation, with a further top-dressing of manure after a harvesting of vines.

Harvest The crop is harvested every four months or can be left for use when extra feed is needed.

Pests and diseases Sweet Potato Weevil causes damage to both tubers and stems. Control by dipping planting material in a solution of Endosulfan. Sweet Potato Mosaic is only controlled by planting resistant varieties.

Hedges and fences

Sisal is a useful hedging material. Mauritius or Lunguzi thorn also makes a good hedge providing it is cut regularly and not allowed to become overgrown. Generally however, barbed wire fences have been found to be most satisfactory, as there is no delay as there is in waiting for the hedge to grow. Also the land is fully utilised, as a hedge takes up quite a considerable area of ground. (See Fig. 1.19.)

If wooden posts are used then wood ash or old engine oil can be poured into the holes. This will help to keep white ants away. The posts should be set in the ground every 6 m to a depth of 60 cm. The posts should have 10–15 cm butts and be 2.25 m long.

Wooden spacers Wooden spacers can be used between posts to keep the wires the right distance apart. If made of wood they should not touch the ground because of white ants. Bamboo makes good spacers, or plain wire can be used. Thin wire can be used to fasten the spacers.

Corner posts Corner posts should be 2.5 m long and strong with ends at least 15 cm wide. They should be set 1 m deep into the ground. They must be braced with poles or with wire.

The fence Four or five strands of barbed wire are usually used. The wires should be slightly closer together at the bottom than at the top to keep in young stock or goats. Use a measuring stick to place the wire at the correct height on each post. The wire should be stretched tightly but do not jerk the wire or overstrain it. When barbed wire breaks, it whips back and can inflict serious cuts. Wire can be strained tight with a wire strainer or by using a pole as a lever (see diagram).

The wire should be stapled on the *inside* of the posts so that when cattle try to get out, they are pushing against the posts and not against the staples.

Cattle grid This is made of poles or galvanised iron pipes. These are set over a shallow pit. Cattle will not cross the grid, but vehicles can drive across. A grid is expensive but useful on larger farms.

Range cattle management

Introduction

There are very large range areas in the tropics which support herds of hardy cattle, sheep and goats, in addition to the donkeys and camels which are even more well adapted to the harsh conditions. Such areas have low stocking rates compared with more fertile agricultural areas.

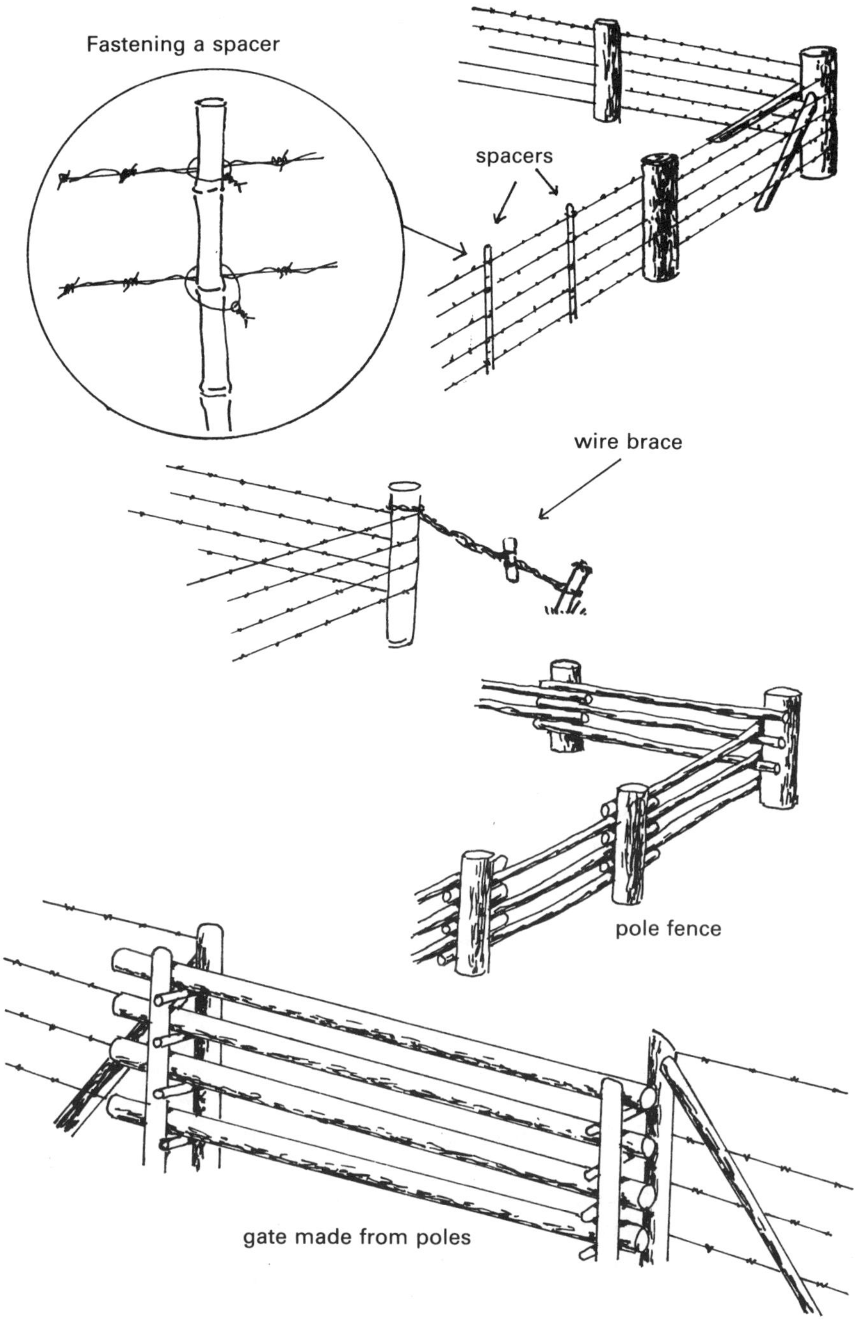

1.19 Fencing and gates

The range areas are the arid and drier lands which are generally unsuitable for arable cropping. They can be described as semi-desert areas, rangeland areas, and ranching areas.

The arid lands and drier rangelands are characterised by generally low elevations, high temperatures, rainfall less than 600 mm/year and high evaporation, due partly to the drying winds. The soils are mostly sandy, and the vegetation is of coarse grasses, thorn bush, acacia bush and baobab trees. Most of these lands are grazed except for areas lacking in water or inhabited by the tsetse fly.

Stocking rates

Stocking rates will vary with the season and availability of grass. An average stocking rate for mature animals is given below for different areas.

Area	*Area per mature animal (ha)*
Semi-arid land	20
Rangeland (fairly dry)	10
Good grazing ranchland	3
Arable land down to pasture	0.5

Semi-desert areas

Pastoralists usually have mixed flocks of cattle, sheep and goats and this is a good practice as the different stock eat different grasses and shrubs.

The pastoralists (often nomadic) in these areas have achieved a balance with nature and a system for survival under extremely difficult conditions. Only the hardy zebu, goats, fat-tailed sheep, donkeys and camels can survive in these arid lands.

There is very little that can be done at present to assist pastoralists in these areas; the ecological balance is so delicate that advice should not be given without a very comprehensive study first being made by rangeland experts.

Rangeland areas

In these fairly dry range areas stock numbers fluctuate considerably. In good years herds build up greatly in number, but when the rains fail many of these animals die. The build-up of stock is an insurance or survival kit for the nomadic pastoralists. The best that can be hoped for in the drier areas of the rangeland is a system of rotational grazing by agreement with all the herdsmen in the area, control of fire, and the sale of some of the surplus stock built up in good years.

Some improved breeding can also take place but the most important characteristic of cattle under these conditions is the ability to survive, and any breeding programme must be made with this in mind.

Some pastoralists keep large herds for prestige and their animals may not be selected because of production potential but because of colour or length of horns, etc. These pastoralists sell cattle only when forced to in times of drought when there is no grazing. At this time, however, every other pastoralist is selling off stock and the prices are therefore very low.

Ranching areas

These ranchlands can be fenced for stock control and it may pay to re-seed the pastures. The most useful grasses in these areas are *Panicum maximum* and *Cenchrus ciliaris* (Guinea grass and African Foxtail). Crops of millet can also be grown in years of good rainfall.

Improved breeding may also be economic as, with good management in these ranching areas, there is likely to be sufficient good grazing to support more productive stock.

Breeding

On the better lands stock can be selected for upgrading. The selected stock are bred with bulls of proven performance under natural or artificial insemination systems. The benefits are:

(a) use of Sahiwal and Boran bulls on the indigenous cattle greatly improves birth weight and growth rate performance of the progeny;

(b) use of exotic bulls, e.g. Friesian or Charolais, on zebu cattle improves the liveweight gains and mature weight, but the progeny is more susceptible to range stress due to harsh conditions. Thus there is need for higher standards of management, e.g. feed supplementation and disease control.

Breeding season The breeding season should be planned in such a way that animals calve down during the time when there is plenty of grazing for the cows and their calves.

Grazing system

On large ranches a convenient and manageable number of cows in a herd is around 100. Ranchers often herd similar cattle together. Thus steers are usually herded together, cows in another group, calves in another and sheep and goats in another.

This system means that at calving time cows with calves can be kept close to the farm buildings. By grazing calves first on pastures, worm infestations are less likely to become serious for them. The order of grazing over a pasture is normally young cattle, mature cattle, then sheep and goats.

Pasture management

The objective of pasture management is to obtain maximum forage, whilst sustaining the vegetation and without causing soil erosion.

Over-grazing can have harmful effects on the pasture when the plants are very young and their root reserves are low, thus making it difficult for the plants to recover. Grazing is best done using a planned rotation, the benefits of which are:

(a) forage plants are only moderately cropped and are thus given a chance to store food for growth;

(b) pastures are rested in rotation and so mature and re-seed themselves;

(c) less palatable plants have to be eaten when animals are not allowed to roam at will over large areas;

(d) good vegetative cover protects water catchment areas;

(e) damage caused to grass and soil by trampling and trekking long distance is reduced.

A minimum of five grazing areas should be marked out. Each area should then be grazed for three months in the year with one area left ungrazed for the whole year, to rest and re-seed.

Night grazing should be allowed if there is no danger from wild animals, as it is difficult for cattle to obtain sufficient feed when grazing only during the day.

Burning

Natural pasture should not be burnt every year. Burning weakens the grasses and destroys the soil cover and the organic matter in it. It thus also greatly reduces the ability of the soil to retain water. Burning should not be done more than once every four years.

Some people burn early in the dry season. This can produce a flush of green grass. The amount of green grass produced is very small and the grasses are weakened and sometimes killed by having been encouraged to grow during the naturally 'dormant' season. Early burning is thus unprofitable.

Bush sometimes encroaches on and spoils natural pastures. Under these circumstances a late burn will keep the encroaching bush and coarse grasses in check. Every few years therefore, it is sometimes useful to burn late. Burning should only be done just before the rains start. This is because grass shoots 6–10 days after burning and if rain does not fall shortly after this, the grass will die.

The burning should be done on a hot day, when there is a moderate wind. This will give a 'hot' burn, because everything is dry, and the fire will then destroy most of the bush. The area should not be grazed for about three months after burning to give the grass a chance to recover.

Firebreaks Firebreaks should be standard management practice to protect pastures being kept for the dry-season feed of cattle. A suggested firebreak is one composed of two bare parallel tracks 4 m wide and 70 m apart. The ground in between should be burnt early in the dry season. One of the strips could be a farm road.

Trees and shrubs

Many trees and shrubs are grazed by cattle and goats. The trees come into leaf before the rains start and they provide good quality food (leaves and pods) when other food is short. Generally, however, the presence of too many trees and shrubs decreases the carrying capacity of land (because there is less grass) and may encourage erosion, because of the poor grass cover under the bush. Bush can be controlled by slashing; however, any trees should be retained for shade.

Chemical bush control Bushes can be sprayed with an 'arboricide' such as Dicloram. After two or three weeks the leaves begin to wilt and the bushes die. The use of an arboricide is expensive and can only be justified on good ranchland.

Erosion

A grass cover should be maintained at all times, by not overgrazing, to prevent erosion and also to retain as much of the rainwater as possible by slowing down rainwater run-off and allowing it to soak into the soil.

Ranchers should not concentrate grazing animals on any one area for too long, as this causes too much trampling. This loosens the dry soil surface and makes it liable to erosion. If additional watering points can be provided, then this cuts down on trekking and cattle concentration.

Gullies can be controlled and further erosion prevented by constructing brush and stone breaks. Sisal can also be planted around gullies as the roots help to retain the soil. Once a gully has been stabilised it can be planted with grass.

Re-seeding

Potentially good ranchland may be improved by re-seeding. The seed may first be mixed with some dry sand to make broadcasting easier; a thorn bush can be dragged over the ground to cover the seed afterwards.

Watering stock

The rancher should provide clean water for his animals. The watering points should be well spaced out to minimise damage to grass by excessive trampling. The drinking troughs should be cleaned daily.

The main source of water in rangelands is rainwater collected in surface dams and pans. This water can be kept clean by ensuring that the soil is not left bare in dam catchment areas through cutting trees and overgrazing. This will prevent silting up of the dams. Also, animals should be encouraged to drink from troughs, and not directly from the dams, as they may pollute the water. Dams should be fenced to prevent this. People should carry water away from the dams to do their washing to avoid polluting the dam water.

Handling facilities

Good, safe handling facilities are particularly important on ranches as some of the cattle are not used to being handled. These facilities include holding yards equipped with crushes. These are used during drenching, dehorning and castration. Crushes should be strongly constructed.

Spread of diseases Stagnant water, swampy areas and irrigated areas are usually infested with certain snails which are the intermediate hosts for liver flukes. The water may also contain leeches which suck blood from animals.

Night security Night yards are used to protect animals from predators and stock rustlers. They can be made from stout poles and barbed wire. Temporary yards are sometimes made from thorn bushes. Minerals and supplementary feed can also be provided in the night yards.

Identification For ranch administration every animal should have a card which has details of name, sex, colour, pattern of marking, breed, etc. The animal should have a mark for field identification. The methods used include branding, ear tattooing, ear-tagging and ear notching.

Castration (See under 'Cattle handling', p. 9.)

Dehorning (See under 'Cattle handling', p. 10.)

Weighing Weighing should be done once every month. The birthweight of calves is also recorded. Weighing helps the rancher to assess the growth rate of the stock and shows which animals have reached slaughter weight. If weighing scales are not available, then weight can be estimated by measuring the girth and length. See under 'Cattle handling', p. 11.

Drenching During the rainy season, regular drenching of the animals against internal parasites such as flukes, tapeworms and roundworms is necessary. Drenching is also necessary for any new animals purchased to prevent them bringing in parasites.

Vaccinations For range animals it is important to follow strict vaccination programmes against foot and mouth, rinderpest, blackquarter, anthrax and brucellosis diseases (see notes on cattle diseases under 'Cattle diseases and control' p. 47). These vaccinations should be undertaken by a livestock officer or animal health assistant.

A full list of regular vaccinations is found in Appendix II.

Wound treatment Wounds should be washed clean with diluted disinfectant or a salt solution. The wound can be covered with a clean cloth. Rest the animal. If the wound is severe seek veterinary advice.

Tsetse fly

In some areas the tsetse fly greatly restricts or even prevents cattle ranching. The tsetse fly feeds by biting the animal and sucking its blood. In the process it transmits blood parasites causing an acute or chronic disease known as trypanosomaiasis.

The flies are found in bushy and humid areas. During the day they rest under shady leaves and become active mainly in the mornings and evenings when temperatures are lower. The fly bite is very painful and on being bitten, the animal jumps or runs away. These areas should not be grazed in the early morning or evening when the flies are more active.

Characteristics of an animal suffering from trypanosomiasis are that it becomes thin, has muscular twitching and will suffer from anaemia. Infected animals should be treated with a trypanocidal drug or they may die.

Control Tsetse fly control methods include clearing the bush and spraying insecticide on the bush. Both of these are expensive activities.

Moving animals

Animals may have to walk long distances from one place to another, in search of grazing land or salt, or to go to market. The effects of making animals walk long distances too quickly are:
(a) animals use much of their strength and thus lose weight and body condition;
(b) if they do not die, the animals get to their destination in very poor condition, and only fetch low prices.

Movement should be done slowly and food and water must be available regularly.

Ticks

Ticks are a major problem in many ranching areas and their control must be included in routine management. See page 49 for full details.

Cattle diseases and control

Agents of disease

The main agents of diseases are bacteria, protozoa, viruses, fungi and parasites. These are all living organisms. The effects of attack by living organisms on an animal are often stronger and more noticeable if the animal is under stress (e.g. if it is suffering from lack of food, subject to excessive cold or physically damaged, for example, by being cut or bruised). In humid and warm conditions these organisms multiply faster.

Bacteria

Bacteria are tiny single-celled organisms and they can be found almost anywhere in nature, many being harmless to livestock. The harmful bacteria, however, are termed 'pathogenic bacteria'. Bacteria are capable of multiplying very quickly which explains the rapid build-up of some diseases in livestock. Bacteria are very sensitive to moisture, light and food requirements. A dry atmosphere will kill bacteria rapidly. Light slows down their growth or even kills them.

Bacteria cause disease through chemicals they produce called 'toxins', which are released into the blood stream, killing tissues and changing chemical reactions in the body from normal to abnormal.

Protozoa

Protozoa are single-celled organisms of the animal kingdom. Unlike bacteria, they have 'tentacles' which can catch food and propel the protozoa, making them mobile. Protozoa can be transmitted to livestock directly, as happens, for example, with coccidia, or by flies and ticks as in trypanosomiasis or the tick-borne diseases.

Viruses

Viruses are tiny particles of protein and are not complete cells. They are able to reproduce only inside other cells. They can cause diseases through the release of chemicals which cause invaded cells to change from their normal function to producing new viruses. This results in tissue damage and disease.

Viruses are able to infect livestock in many ways: through close contact with other infected animals, through infected food, or transmitted via other organisms such as mosquitoes or (as in the case of rabies) by dogs and other animals.

Viruses are able to withstand many hazards; however, nearly all viruses are killed by fairly moderate heat.

Fungi

The fungi belong to the plant kingdom. They are unable to manufacture their own food and so use products from other plants or animals, both living and dead. They feed by releasing chemicals which dissolve the food substance, and then they absorb the solution. Some fungi feed off the skin or hide whilst others live inside the animal: for example, in the mouth, stomach or gut. Fungal diseases tend to spread slowly.

Parasites

There are several organisms which live on or in livestock which are called parasites. Parasites may affect the host animal in a variety of ways:

(a) by taking the host's food, e.g. tapeworms in the gut;
(b) by sucking blood, e.g. ticks and roundworms;
(c) by causing damage to some organs, e.g. the liver fluke which eventually destroys the host's liver.

Worms

There are three types of worm affecting livestock: tapeworms, flukes and roundworms. Worms generally live inside the animal, eating either its food or the animal itself. The symptoms produced are poor condition generally and emaciation (thinness), with the ribs and backbone showing prominently; the belly may also be swollen.

External parasites

Organisms which live on the outside (e.g. on the skin or in the hair) of an animal are lice, fleas, ticks and mites. These either feed on the skin itself, or suck the animal's blood. They are important because of the local irritation they cause and the damage to the hide, but most of all because they transmit a number of serious diseases to the stock.

Disease prevention

Disease can largely be prevented by cleanliness in the animal's environment, preventing access by possible diseases-carriers and killing host insects.

Environmental control The housing, feeding and watering equipment should be cleaned regularly to prevent them being a source of disease.

Movement control A farmer's animals should be isolated as far as possible from contact with other animals as these may be sources of disease. In addition, visitors should not be permitted close to the stock or in buildings as they may carry disease on their boots or clothes.

Control of host insects Where diseases are carried by insects such as ticks or flies, then a programme to kill such insects is necessary.

Control by vaccination There are certain diseases which can be adequately controlled only by vaccination.

Veterinary attention Sick animals should be isolated as soon as possible from the rest of the herd and should receive urgent veterinary attention. Many diseases can be cured if treated at the early stage.

Tick control

The two most important health measures are tick control and vaccination against certain diseases of major importance. Tick control, if done properly, will prevent the whole group of tick-borne diseases which are transmitted from affected animals to healthy ones by ticks. These diseases are East Coast fever, corridor disease, anaplasmosis, redwater and heartwater.

Furthermore, ticks are blood-sucking parasites and cause great loss in production by weakening the animals through loss of blood, irritation, infections starting in the open wounds left by the ticks, and damage to hides.

Tick cycle

After hatching from the eggs, tick larvae climb up vegetation. They then attach themselves to any passing animal. The larva sucks blood from the host and injects a fluid to facilitate the flow of blood. It is this saliva or fluid which may carry the tick-borne disease parasites. The larva becomes a nymph and after further blood sucking and moulting, becomes an adult tick. Adults then mate and drop off the host for the female to find a humid, sheltered place to lay her eggs.

It is important to kill the ticks before the females can lay their eggs. As ticks breed at different times it is necessary to keep up a spray or dip programme continuously.

Tick control can be carried out by dipping or spraying, the most common method being dipping. In addition to these, 'hand dressing' may be done.

Dipping

In order to run a dip efficiently certain points must be strictly followed.

Cattle handling Dip only in the early morning and not when it is raining. Move the cattle to the dip slowly and allow them to drink water and rest before dipping them.

Work the cattle through the dip slowly and quietly.

Do not dip sick animals or cows close to calving.

Strength of dip The dip wash must always be kept at the correct strength. As cattle are dipped, the dip loses strength, so the dip attendant should keep a dip register, showing the number of animals dipped and the amount of dip chemical used. For the replenishment rate he will have to follow strictly the instructions given by the manufacturer of the chemical.

A dip sample should be sent for testing once or twice every month.

The dip must be kept clean: a dirty dip, even if it contains enough chemical, will not kill ticks effectively. The farmer should check on the dip attendant to make sure that he does his job properly. If the dip solution is not kept up to strength then the ticks will not be killed and farmers will see no purpose in taking their cattle to the dip.

Spraying

The spray chemical must be mixed exactly according to the instructions given by the manufacturer. If hand spraying is done, the animals should be sprayed thoroughly all over, particularly in

(a)

(b)

Take special care to spray in the concealed parts of the animal.

1.20 Cattle spraying:
(a) spray race, (b) hand spraying

the ears and under the tail. If a spray race is used, all nozzles must be in good condition and the pump pressure must be correct so that 700 litres of spray are delivered per minute. (See Fig. 1.20.)

After use, flush all pipes and nozzles with clean water to prevent a build-up of oily residue.

Hand spraying Animals which cannot be dipped can be hand sprayed. This is time-consuming and so should be done only when small numbers of animals are involved. The same chemical concentration as for the spray race or dipping bath should be used.

In order to wet the animal thoroughly you should apply at least 10 litres of dip wash with a high-pressure spray pump.

Hand dressing

Spraying or dipping may be accompanied by dressing with an anti-tick grease. A piece of cloth or a brush should be moistened with the grease and smeared into and around the ears and under the neck and tail and on to the tail itself, these being the favourite feeding placed of the brown ear tick which transmits East Coast fever.

To make hand dressing and spraying more efficient, the hair inside the ears and also the tail brush should be clipped. This can be done every six months.

Dipping and spraying intervals

All animals, whether grade or indigenous, must be treated at least once each week. During wet periods, which suit ticks and therefore increase their population, it is advisable to do it at five–day intervals or even twice every week in heavily infested areas.

Table 1.10 Types of dip chemicals (acaricides)

Chemical	*Manufacturer*	*Trade name*
Carbaryl	Murphy	Sevin
Chlorfenvinphos	Shell	Supona
Coumaphos	Bayer	Asuntol
Dioxathion	Wellcome	Delvav
Dioxathion and Chlorfenvinphos	Wellcome	Supamix
Oxiniothiophos	Bayer	Bacdip
Camphachlor	Wellcome	Coopertox
	Pfizer	Pfizertox

Classification of diseases

Livestock diseases and ill health can be classified under five headings: tick-borne diseases; other bacterial and viral diseases; parasitic diseases; nutritional diseases; breeding diseases.

The majority of the diseases described below are serious; it is strongly advised that the farmer consults his veterinarian when his animals are sick.

Notifiable diseases

Certain diseases are 'notifiable diseases'. They are either dangerous to man (e.g. anthrax) or they can cause very big economic losses. In most countries these are notifiable to the veterinary authorities; that means cases or suspected cases of these diseases must be reported to the nearest veterinary office without delay. Failure to do so can make the owner of the animal liable to prosecution.

Tick-borne diseases

Anaplasmosis

Caused by a protozoan parasite and transmitted by a number of ticks, the most common being the blue tick.

Symptoms Rise in animal's temperature, dullness, listlessness, loss of condition. Constipation is characteristic; anaemia, and in the later stages, jaundice, seen in the eyelids, gums and vulva.

Treatment Tetracyclines can cure the disease if they are given in time, but prevention by tick control methods is both safer and cheaper.

Redwater (Babesiasis)

Caused by a protozoan parasite transmitted mainly by the blue tick.

Symptoms Rise in animal's temperature, dullness, lack of appetite; the visible mucous membranes show anaemia; breathing rate increases; the urine can become reddish-brown in colour and foamy. Death can occur within a very short time of the onset of the disease.

Treatment Prescribed drugs can cure the disease provided they are given before the animal has become too anaemic.

East Coast fever

Caused by protozoan parasites which are transmitted by the brown ear tick.

Symptoms Increase of animal's temperature up to 106°F. The animal is dull and listless. Swelling of the lymph glands below the ears, in front of the shoulders or in front of the knee joint are usually to be seen. In advanced stages, respiration becomes laboured, the animal coughs, it loses condition, becomes weak and eventually is unable to rise. Froth usually appears from the nostrils when it dies.

Treatment No treatment is available. Therefore prevention by tick control is of the greatest importance.

Heartwater

Caused by a protozoan parasite which is transmitted by amblyomma ticks.

Symptoms Rise in animal's temperature, nervous symptoms such as champing of the jaws and muscle twitching. The animal may walk in circles or against obstacles. Once lying down, it may paddle its legs before dying.

Treatment Treatment with sulphonamides or tetracyclines is effective if given in the early stages. Prevention by proper tick control is both safer and cheaper.

Bacterial and viral disease

Blackquarter

Caused by a bacterium; it is very similar to blackleg.

Symptoms Rise of temperature, loss of appetite, lameness, swelling of hip or shoulder which feels crackly due to gas formation.

Treatment Penicillin is sometimes effective if given in the very early stages. Dead animals should be burnt not buried.

Prevention As for anthrax.

Anthrax

Anthrax is a bacterial disease, very dangerous to cattle and also to man.

Symptoms High fever followed by rapid bowel inflammation and death. After death, blood oozes from the anus, nose and mouth. Often the disease kills so fast that no symptoms are seen, the animals are just found dead (in all such cases anthrax has to be suspected).

Treatment Penicillin is effective if given in the very early stages.

Precautions A dead animal suspected of anthrax infection must not be opened up. Anthrax bacilli which get in contact with air will form what are called 'spores'. These spores can survive in the soil for more than 20 years and re-infect animals or man. Therefore the animal must be buried deeply in a dry place, together with any soil contaminated with excrement or blood of the animal; alternatively, it is better if the carcass is burned.

Prevention Effective vaccines are available which produce an immunity against blackquarter, as well as anthrax. Animals should be vaccinated once every year, or in high risk areas, once every six months.

Contagious bovine pleuro-pneumonia

Spread by breathing in droplets discharged from the noses of other infected cattle.

Symptoms Rise in temperature, dry, staring coat, followed by hard, painful coughing and discharge of thick mucous at the nose and mouth.

Treatment None.

Control Slaughter of infected animals, quarantine and vaccination of herd.

Brucellosis

Caused by bacterium. The usual way of infection is by mouth, the main source being pastures contaminated by an aborted foetus, afterbirth or vaginal discharge from a cow which has aborted. Brucellosis is also known as epizootic abortion, undulant fever or Bang's disease.

Symptoms Abortion, usually after six or seven months of pregnancy. In male animals the scrotum and testes become enlarged and painful.

Treatment Treatment is unreliable and uneconomical.

Control If any suspicious abortion occurs, the foetus or part of the placenta should be sent to a veterinary laboratory for confirmation. If this is not possible a blood sample should be taken from the dam about three weeks after abortion and sent for testing.

Suspected animals should be kept separate and if the disease is confirmed, should be culled. If culling cannot be done because too many animals are affected, then all young stock should be vaccinated at the age of six to eight months. Vaccine is available from veterinary departments.

Precautions Male animals which have been infected with the disease should not be used for breeding.

Brucellosis can affect man, the main ways of infection being either by the handling of an aborted foetus or a placenta when having scratches or small wounds on the hands, or by drinking unboiled milk from an infected cow. The main sympton is high fever which lasts about a week and then returns after a period of up to three months.

Foot and mouth disease

Foot and mouth disease is an extremely contagious, acute disease which can affect all cloven-footed animals. It is caused by a virus. The disease spreads rapidly and usually very many animals are affected within a short time.

Symptoms The disease is characterised by high fever, salivation and lameness caused by blisters in the mouth and on the feet. Unlike East Coast fever, foot and mouth disease is not a killing disease, but the disease is of paramount importance because of its economic effects, namely severe loss of body weight in beef animals.

In dairy cattle, milk yield can be cut by 60 per cent or more and is not recovered in that lactation. Pregnant cows can abort and it usually takes a long time to get them in calf again.

Complications by bacterial secondary infections can lead to mastitis and chronic foot infections which can cause permanent lameness and emaciation.

Treatment There is no known specific treatment. Antibiotics can help only against bacterial secondary infections.

Prevention Vaccination and livestock movement control are the only ways of getting foot and mouth disease under control. Because of the importance of this disease in many countries, all cattle from the age of four weeks living in areas of high risk from the disease must be presented for vaccination every six months and must not be moved from one area to another without permission. Failure in either case can make the owner liable to prosecution.

Unfortunately one vaccination against foot and mouth disease does not produce a very high or long-lasting level of immunity. Therefore it is necessary to boost this immunity by revaccination every six months in order to prevent the animals from catching the disease.

Rinderpest

Rinderpest is another dangerous virus disease but fortunately it is decreasing in importance in places where vaccination programmes have been carried out for many years.

Symptoms Rise in temperature; a harsh, dry coat, then discharge from the eyes and nose; mucosal lesions in the vulva, vagina and in the mouth; laboured and painful respiration; severe diarrhoea which causes rapid dehydration and death.

Treatment No treatment known.

Prevention All animals over one year of age must be vaccinated. Fortunately one vaccination produces an immunity which lasts for the animal's lifetime.

Johne's disease

Caused by a bacterium. Infection is passed on by mouth.

Symptoms Profuse, persistent, smelly diarrhoea, resisting any treatment. The animal loses condition and eventually dies. Animals below the age of four years are rarely affected.

Treatment No specific treatment known. Affected animals should be removed and slaughtered before they can infect others.

Lumpy skin disease

This is an infectious viral disease transmitted, it is believed, by mosquitoes and other biting flies e.g. stomoxys.

Symptoms Rise in temperature, loss of appetite, salivation and a clear discharge from the nose; later followed by the appearance of lumps in the skin.

Treatment Secondary, bacterial infections should be treated with penicillin or sulphonamides.

Control Vaccination.

Mucosal diseases complex (virus diarrhoea)

These are viral diseases having many symptoms similar to foot and mouth disease, and are important for that reason.

Symptoms High fever and pulse, lack of appetite; profuse, smelly diarrhoea which may contain blood or mucous; ulcers on the lips and gums.

Treatment None.

Control None.

Rabies

A highly fatal, viral disease affecting all mammals, including man. It is normally transmitted through a bite from an infected animal.

Symptoms In cattle, uncoordinated movement; hind legs tend to collapse; paralysis of throat preventing eating or drinking. Alternatively the animal may madly attack objects or other animals. Death occurs in a very few days.

Treatment None.

Control Vaccination, particularly of dogs.

Trypanosomiasis

Caused by a micro-organism transmitted through the bite of tsetse flies.

Symptoms In acute cases there is high temperature, followed by anaemia, progressive weakness and death. In chronic cases the temperature varies, the animals have dry coat, are listless and become very thin. Death usually occurs after about three months.

Treatment Several drugs are available on prescription from veterinary departments. The injection of the drug cures the disease and protects the animal for about two months.

Prevention Extensive bush clearing to destroy the tsetse flies' living places.

Tuberculosis

A serious disease of both cattle and humans, caused by micro-organisms of the mycobacterium group.

Symptoms General emaciation; low, moist coughing, especially after exercise. In cases of udder infection, small lumps can be felt in the udder which also becomes enlarged. The milk curdles, becomes thin and green-coloured.

Treatment Treatment is available, but it is generally both safer and cheaper to slaughter infected animals.

Control Culling of animals not resistant to the disease as shown by a tuberculin test.

Mastitis

A bacterial infection of the udder causing serious loss of milk yield, permanent damage to the animal or even death. Caused by improper milking or bruising to the udder.

Symptoms Inflammation of one or more quarters of the udder. Usually it is painful to the cow, so the first sign is often when she kicks the milker. The udder becomes enlarged and hard. The quality of the milk changes, becomes flaky, may be streaked with red, or yellowish in colour. Yield is severely reduced and may never be recovered after treatment.

Treatment Treatment must be started without delay. The affected quarter of the udder is emptied of milk, filled with an antibiotic (see Fig. 1.21) and left for about twelve hours. The treatment is repeated as necessary. If the cow is not eating and her temperture is above normal, an antibiotic injection should also be given.

1.21 Administering antibiotic to cow for control of mastitis

Precautions Animals with mastitis should be milked last. The milk should not be mixed with the milk from the rest of the herd whilst the udder is receiving antibiotic treatment and for at least three days after successful control of the disease.

Prevention Cleanliness in milking and using the correct method. Use disinfectant when washing the udder before milking and a teat dip afterwards. Machine milking equipment must be scrupulously clean and correctly adjusted.

Pink eye

Several infectious bacterial and viral agents are thought to be responsible for this complaint which tends to occur in dry, dusty environmental conditions. Blindness may result if treatment is delayed.

Symptoms Tears in the eyes, conjunctivitis and varying degrees of cloudiness and ulceration of the eyes. Animals become sensitive to the light and prefer to remain in dark places.

Treatment Antibiotic eye ointment or powder is effective, provided treatment is started in the early stages. Keep the animal in a darkened place and protect from flies.

Three-day sickness

Three-day sickness is a virus disease which is transmitted by culicoides insects and consequently occurs seasonally at the beginning of the rains when the population of these biting insects goes up.

Symptoms Transient fever, shifting, muscular pain and lameness.

Treatment No specific treatment. Affected animals usually recover after three to four days.
Note: Animals suffering from three-day sickness must not be drenched because the disease sometimes causes paralysis of the throat; the animal is not able to swallow properly and so the drench might get into the lungs.

Parasitic diseases

Internal parasites

The more intensive the grazing, the greater the likelihood of a heavy build-up of internal parasites. Adults develop some resistance but calves can suffer badly. Even moderate infestations can reduce production considerably, though high feeding levels can mask the effects of the worm infestation.

Complete elimination of the usual parasites is not practical and may not even be desirable as other more dangerous parasites may then build up. The relationship of worms to stock is complicated and not yet fully understood.

Rotation of pastures keeps the worm population down to reasonable levels. Routine worm control measures are essential in tropical countries as the warm, humid conditions are ideal for worms.

Symptoms of worms Failure to thrive, anaemia, weakness, loss of appetite, diarrhoea. Lung worms may cause coughing and an increased rate of breathing. In heavy infestations, calves may have to be treated every six weeks. Liver fluke is difficult to diagnose except from a stool sample or slaughtered animal. The best prevention is to keep cattle out of wet stream areas and treat routinely against the flukes every three months.

Method of dosing for internal parasites Restrain the animal so that it cannot jump around. The correct amount of medicine is measured into the bottle or gun.

Keep a hand over the animal's nose and use the fingers to open the mouth and keep back the head. Never hold the tongue out of the way. Open the mouth just enough to admit the gun or bottle neck. You may be severely bitten if your fingers are inserted between the molar teeth.

Place the bottle neck in the corner of the animal's mouth. Ensure that each small quantity of medicine is swallowed before the next is given. This is especially important for calves, otherwise they may choke.

After dosing, hold the mouth closed for a few seconds after withdrawing the gun or the empty bottle, so that you are sure the dose has been completely swallowed.

Timing A dosing schedule is given on page 131.

Nutritional diseases

Bloat

Bloat usually occurs after a dry period, when animals are put on young, fresh grass which contains a lot of protein but not enough roughage and therefore produces too much gas in the rumen. The rumen then expands rapidly, putting such pressure on the lungs and heart that the animal dies.

Treatment Commerical bloat remedies are available to be given as a drench. At the same time the animal should be walked briskly.

Milk fever

Milk fever usually occurs one to three days after calving, but can also occur beforehand. The most common sign is the cow lying in the normal resting position but with the head and neck bent sideways; the cow appears to be asleep.

The sudden onset of milk production lowers the blood calcium level which causes paralysis, unconsciousness and eventually death.

Treatment Treatment is urgent. Calcium borogluconate solution is injected intravenously or subcutaneously. Contact a vetinarian as quickly as possible. If a vetinarian is not immediately available then subcutaneous injections in a number of sites should be administered within a few minutes, rubbing the area gently by hand to encourage absorption of the calcium solution. It is often the highest yielders which are affected and calcium borogluconate should be kept as a standard item on all dairy farms.

Breeding diseases

Leptospirosis

A highly contagious disease affecting all ages of cattle but more severe for calves. It causes abortion is pregnant cows.

Symptoms The symptoms are easily confused with those of redwater. Calves show fever, prostration, loss of appetite, difficulty in breathing, yellowing of mucous membranes, red urine and anaemia.

In older cattle the symptoms are milder and more variable. The fever period is short and accompanied by loss of appetite and depression. In milking cows there is a sudden drop in milk production; the milk is thick, yellow and blood-tinged, although the udder shows no abnormality.

Abortion is a common feature in pregnant females and is commonest about the seventh month of pregnancy. An abortion outbreak in a breeding herd is often the first indication of leptospirosis.

Treatment Antibiotics are effective if given early enough.

Control Attempt to identify carriers and give vaccination to susceptible cattle. Do not mix pigs with cattle as pigs are well-established natural reservoirs of infection. Other animals affected are sheep, horses and dogs, as well as humans.

Trichomoniasis

A venereal disease caused by a protozoan and transmitted sexually.

Symptoms The disease is characterised by abortion (usually after three to four months of pregnancy), discharge of pus from the uterus and sterility.

Treatment Washing out the uterus with Lugol solution cures the infection. To avoid re-infection by bulls (which are very difficult to treat successfully) use artificial insemination.

Vibriosis

A venereal disease.

Symptoms Early death of the foetus, infertility or abortion after two to four months pregnancy.

Treatment If re-infection is avoided by using artificial insemination, infected cows will recover without treatment after one or more heats. The infection also responds to antibiotics.

2 Donkeys

Management

The donkey is a hardy animal and extremely strong for its size. It is used the world over as a pack animal, carrying loads on its back or pulling carts. The donkey is one of the most efficient power units in agriculture, and under difficult, hilly conditions or on small farms, is much more economical than a mechanical unit. Donkeys are sometimes misused or badly looked after. In these cases they become unfit for work.

Breeding

The gestation period of the donkey is about 362 days, and only one foal is born at a time. Special care should be given to the donkey when she is pregnant. A pregnant donkey should not be worked for the last two months of pregnancy and until the foal is weaned, about three months after birth.

A female donkey should not be allowed to breed until she is three to four years of age so that she can grow to a good size.

Feeding

Donkeys often seem to do well on just grazing. However bran may be given, as well as vegetables such as carrots and cabbage, if they are available. Hay and maize trash may also be given, particularly when grass is short during the dry season. Pregnant donkeys should receive good grazing or hay, and 0.5 kg concentrates daily for two months before and three months after birth.

Clean water should be available for the donkey several times each day. Time should be found during work periods for watering, particularly on hot days.

During weaning and for several months afterwards, the foal should be fed 0.5 kg of concentrates per day. (Ground maize, wheat, beans, cotton-seed cake; see under 'Dairy cattle management'.)

Housing

A simple shelter should be provided to protect the donkey from the sun and rain. If it is necessary to tether the donkey, use a stout rope about 5–10 metres in length, tied round a leg, to allow extensive grazing. It should be firmly tied, but not so tight as to cause sores.

Working conditions

Donkeys should not be overworked; they should work for not more than two or three hours at a time. Care should be taken that carts or loads are not too heavy otherwise the donkey will be injured. It is preferable that the donkey pulls a cart by means of a harness distributing the weight over the neck, back and chest, and not just the neck, (see Fig. 2.1).

Harness

A good harness can be made from sheepskin (Fig. 2.2) but if this is not available, thick, flat webbing will be sufficient. Rope should not be used alone as this can cut into the flesh and cause sores. However, if it is kept away from the skin by suitable material such as several layers of sacking, it may be used.

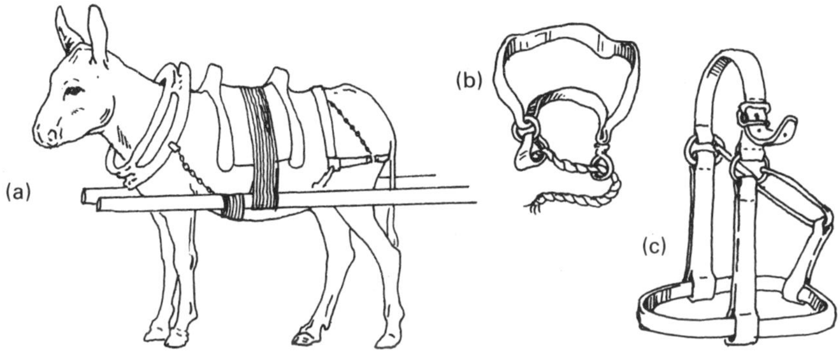

2.1 Donkey harness:
(a) for pulling cart, (b) halter, (c) head collar

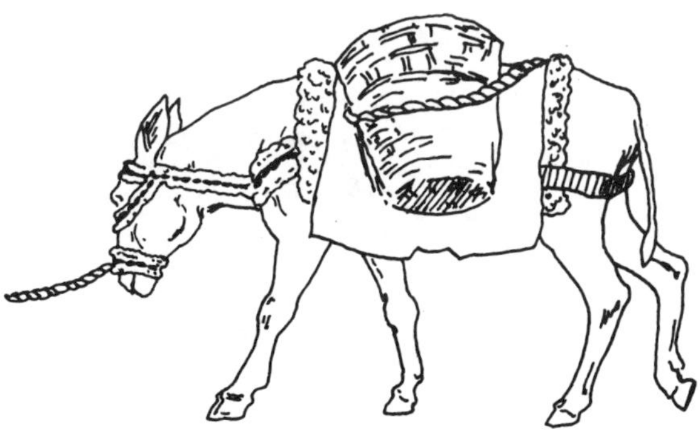

2.2 Laden donkey showing padded harness

The harness must be renewed regularly. Those made from sacking soon become very stiff with dirt and sweat and continued use will result in sores. The fur on sheepskin will wear off and this then also causes sores. Should a sore develop, it should be treated daily with Stockholm tar. The donkey should not be worked until the wound has healed. The harness which caused the sore must be renewed.

Carts

Carts should be as light as possible, as with a heavy cart only small loads can be pulled. Shafts should be set well apart to prevent rubbing against the donkey's sides, and long enough so that the cart does not bump against the donkey's legs, particularly when travelling downhill. A cart must have a handbrake to assist in slowing down when going downhill and to hold the cart stationary when necessary.

When more than one donkey is used, suitable harnessing should be used for the side donkeys. Harnessing one donkey behind the other is a good method.

Never use a whip or heavy stick when driving a donkey; a light switch together with commands or whistles is quite enough.

Care and maintenance

Foot care

The most important aspect of caring for a donkey is keeping its hooves clean and well trimmed. All donkey owners should be in possession of a hoof pick (even a bent six inch nail will do) to keep the frog of the hoof free from dirt and mud which will quickly cause thrush, an infection easily detected by its unpleasant smell. Small stones should also be removed as they can cause pain and discomfort. Great care should be taken whilst 'picking' the foot as it can easily be injured. Hooves should be cleaned daily if possible.

It is also necessary to have the hooves cut regularly (Fig. 2.3) to prevent them becoming overgrown, a condition which eventually

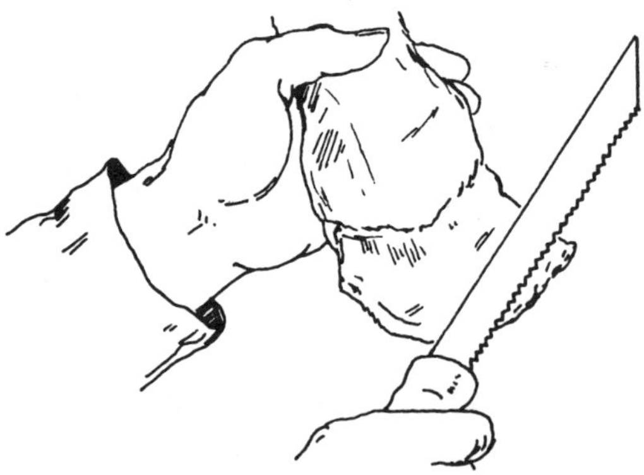

2.3 Use a sharp saw-toothed blade to trim overgrown hooves

causes lameness and much pain to the animal. This must be carefully done by a skilled person. Great care must be taken, as to trim the hooves too much is as bad as not trimming them at all. When in doubt, take off less rather than more. It is advisable to trim hooves every six to eight weeks, as if the donkey becomes lame, it cannot work. Use a good knife and a rasp file.

Coat care

Coats should be looked after. A stiff brush can be used to groom the donkey weekly. This will remove mud and keep the coat healthy and shining.

Worm control

Donkeys should be wormed regularly every six months with a worm medicine used for horses. The signs that a donkey has worms are: general lethargy, staring coat, large belly and visible ribs, as well as segments of worm in the stool.

Donkeys can get mango worm. This can be seen as a boil-like swelling. When the head of the maggot appears it is 'ripe' to treat. It can be squeezed out and destroyed, and the resulting hole treated with an antiseptic.

Eye infections

Donkeys frequently suffer from eye infections. These can be treated with Golden Eye ointment, or with antibiotic cream as prescribed by a vet.

Diseases

Biliary fever

Biliary fever is transmitted by ticks.

Symptoms Rise in temperature, weakness, staring coat and diarrhoea; the urine becomes dark brown in colour.

Treatment Treatment is very effective, but should only be administered by a vet.

Surra

This name is given to trypanosomiasis in horses and donkeys. See 'Cattle Diseases'.

Glanders

A contagious disease, spread by contact with other infected animals. The disease is very dangerous as it can be passed to humans and is fatal.

Symptoms High fever, coughing, ulcers in the nose with a thick, sticky discharge. The glands under the jaw swell up. Death may occur within a few days.

Treatment None.

Control Destruction of infected animals to prevent spread of disease.

Epizootic lymphrangitis

A contagious disease caused by a fungus.

Symptoms Lumps under the skin (in the lymphatic vessels) which break and discharge a thick yellow pus.

Treatment None. Recovery often occurs unaided.

Control Isolation of infected animals; disinfection of harness, stable, etc.

Mange

Skin disease caused by a mite.

Symptoms Scratching and itching, loss of hair, crusty wounds on the skin.

Treatment Dip or spray, as for cattle ticks.

3 Goats

Introduction

Goats have been important domestic animals in many countries for thousands of years. Different breeds have developed for various purposes and to suit different climates and conditions.

Considerable emphasis is now being placed on goats for milk production. This is particularly important in those tropical areas where the human population is large and land cannot be spared for milk cows.

Most goats however live in areas of low potential for arable agriculture or in the semi-arid areas. In these areas the management practices need to be quite different. These goats are kept for meat to be eaten at celebrations or sold or bartered in times of need. The goats are sometimes milked but the milk yield is very low.

Examples of some different goat breeds are given below.

Breeds of goats

Meat Breeds

East African

This is a small, very hardy goat, the adult female weighing around 25 kg. It is kept for meat production, but is sometimes milked, especially by pastoralists. There are various types of East African goat including the Galla, Somali and the Turkana/Samburu, which is long-haired and bearded. The East African goat has some resistance to trypanosomiasis.

Boer

This goat from South Africa is used for cross-breeding to increase meat size and productivity. It is white, with long ears and hair.

Tropical Milk Breeds

Anglo-Nubian

A goat of mixed African and Eastern origins. The colour may be a mixture of roan and white. It has long legs and a Roman curved nose and long pendulous (hanging-down) ears. This breed yields less than Toggenburg or Saanen, but the milk has a higher butterfat content. It is a large strong goat weighing around 60–75 kg. The Anglo-Nubian is also kept in temperate areas.

3.1 Goat Breeds

Jamnapari

This goat is a cross between the Indian Jamnapari and the Egyptian Nubian. It may be white, black or fawn in colour. The ears are large and lopped. Females weigh 45–60 kg. They are horned and may produce some 1.0–1.5 litres of milk per day. This breed originates from India.

Temperate Milk Breeds

Toggenburg

This is a small Swiss breed coloured white and fawn with cream stripes. It has erect forward pointed ears and is usually polled. The females weigh around 40–50 kg.

Saanen

The Saanen is a small, white or pale fawn, Swiss breed. Ears are erect and point forward. It is usually polled and weighs around 50–60 kg.

Management of goats for meat production

Breeding

Under the normal extensive system goats may produce two kids each year or three kids every two years. The percentage of twins is usually less than 10% and in arid regions this is desirable as rearing conditions are very hard.

A female will be ready to mate at five to six months old. This should be prevented with an anti-mating apron worn by the male. It is best not to breed the female before she is 1–1½ years old, otherwise her growth will be stunted.

Males not wanted for breeding should be castrated before four months of age with a Burdizzo or within 24 hours of birth if using an Elastrator. However, this is not necessary if they are to be slaughtered whilst young.

Grazing

Meat breeds are extremely hardy and require little attention. In the drier areas, goats are herded over extensive grazing lands.

With extensive grazing it is not always possible to return home every night and so temporary thorn camps have to be used. The very young kids should not be expected to walk long distances so the mothers should return at midday so that the kids can suckle.

The ranch or range land should be divided up into paddocks or areas, and each one grazed for two to three months. Ideally, one area should be left ungrazed for the whole year so that it can rest and reseed itself. (See grazing section under 'Grazing, fodder crops and fencing', on page 28.)

The proportion of browse and grasses eaten may vary. In the dry season goats tend to eat more browse. They particularly like Acacia pods and will also eat bark and roots. Thus goats are very useful on a ranch in keeping down bush regrowth. In dry areas some ten goats can be kept instead of one bovine animal.

Control of Animals

Where a number of goats are kept it may be useful to have a crush where goats can be controlled for drenching, vaccinations and foot trimming.

If it is wished to dehorn goats, this should be done, by a trained stockman, during the first one to three months of age.

Disbudding should be done during the first four days of life, only by a trained stockman.

Dipping

If a cattle dip is not available, a small dip can be constructed. A dip is necessary as goats can act as carriers of ticks and parasites.

Pests and Diseases

Generally goats are healthy animals. (see disease section under 'Sheep' on page 120). The following additional points on health are made:

Internal worms Dose against internal worms twice a year at the onset of the rains. Graze rotationally to reduce the risk of worm infection.

Ticks Dip or hand pick and hand spray.

Vaccinations Ask the Veterinary Officer to vaccinate against anthrax, foot and mouth and tetanus.

Management of goats for milk production

Most of the goats in the Tropics are meat breeds. In some areas there seems to be a reluctance to drink goats' milk instead of cows' milk, even though it is of equal, if not better, nutritional value as the fat and protein in goats' milk are digested more easily.

With the increasing human population, especially in high potential areas, there is an excellent case for a greatly increased number of milk goats. A family can keep a goat on small pieces of waste or steep ground and feed it crop residues. Five or six milk goats can be kept on the same amount of feed and grazing as required by one cow. One goat will provide sufficient milk for a family. However, it must be realised that dairy goats require the same careful attention as dairy cows, and the temperate breeds need extra care and shade during hot weather. Farmers who cannot provide good management may be advised to keep cross-bred goats, rather than the pure-bred varieties.

Points of a productive milking goat

A good milking goat should have deep wide-sprung ribs, a long sloping rump, a large elastic udder and teats and large milk veins under the belly.

Breeding

Breeding should take place five months before the start of the rains so that the goats will kid at the beginning of the rains when food supplies are good and the mothers will be able to provide plenty of milk.

One male can serve 30–40 females. The service should take place when it is detected that the female is on heat, with a second service 12–24 hours later.

There may be a false heat and if so the female will come on heat 10 days later. Normally there are 18–21 days between heats. Heat signs are:

(a) Excitable behaviour and bleating.

(b) Redness around the vulva and sometimes a mucous discharge.

(c) Frequent switching of the tail.

Heat normally lasts 24–36 hours but sometimes it only lasts a few hours so the female goat should be taken to the male goat as soon as possible after heat signs are observed.

The gestation period is 150 days.

Females should be bred around 18 months of age, not before as they will not be sufficiently well grown.

Management of pregnant goats

The pregnant goat should be dried-off six to eight weeks before she is due to kid. This can be done by greatly reducing the concentrate and by first milking only once a day instead of twice. After another week or so, do not feed any concentrates and milk only every other day.

After she has dried-off, the goat should receive extra feed such as 0.5 kg concentrates per day for some three to four weeks before kidding and, in addition, some succulents such as kale. However, she should not be fed concentrates for four days after kidding because doing so could lead to milk fever.

The goat should be put in a pen about one week before kidding.

Care of the kid

Kids can be bucket-fed, fed on milk substitute, bottle-fed or fed by the mother.

For the ordinary goat-keeper, the last method is the best as it is the simplest and one most likely to keep the kid healthy. However, care needs to be taken as there are certain problems. Firstly, the kid might only suck on one side of the goat. The goat should be milked out twice each day in any case. Secondly, the kid may be difficult to wean. The best method is to shut the kid away at night after it is one month old. The goat can then be almost fully milked out each morning, to provide milk for the family. The kid should then be allowed to suckle for a short period twice each day until it is three to four months old.

When the kids are suckling, the goats' teats may become sore. A little Vaseline (white petroleum jelly) should be applied.

Kids in a pen may like to nibble on fresh earth to get minerals. This can be provided in a box and changed every one to two days. At about two weeks of age a small bundle of good quality hay should be available for the young kids. At three to four weeks offer some concentrates in a box. As they grow bigger they will eat up to around 0.25 kg concentrates per day.

Care of young goats

Young goats should be fed hay and allowed to graze at about six weeks of age. Young goats should graze only on clean land and they should not follow sheep because of the risk of parasite infection. They should be given some 0.25 kg of concentrates per day. If kept in a stall they will require 0.4 kg of concentrates per day.

After mating (at around 18 months) the concentrates ration should be increased gradually to 1 kg. Young female goats need not be segregated and can be left together in a communal pen.

Care of male goats

It is not necessary for every goat owner to keep a male goat. Females when on heat can be taken to a nearby goat owner who keeps a good male goat. A male goat should be housed some distance from milking goats, so that no smell is passed to the milk.

Male kids can be kept for meat production. They should be castrated in the first week and killed at around three months of age.

Feeding mature goats

Hay Hay is a very important food for goats. This provides the bulk and the fibre which is essential in their diet. Goats should have hay provided twice each day. They can be very wasteful with hay, pulling it down and trampling on it. Hay racks can be made which should be 60 cm deep and have slats 4–5 cm apart which prevents the goat pulling out too much hay.

Green feed If the goat is to be given very green feed or put on a lush green pasture to graze, hay should be given first. This will prevent bloat which may otherwise occur. The rich grass can cause the stomach to swell up very badly.

An excess of green feed may taint the milk. However, some green feed is important and should be provided. Suitable green feed includes good quality grass, leafy branches, kale, cabbage, silage, sweet potato and vines.

Water The watering bucket can be located outside the pen with a hole in the wall for the goat to put its head through to drink. This prevents the goat upsetting the water container or fouling the water. The water must always be clean.

Grazing Goats may be allowed to graze in a field. However they tend to break out of the field it it is not well fenced. Barbed wire however should not be used for fencing as this can cause serious injury to udders. 0.25 hectare of land is enough for two goats. This area can be divided into three and the grazing rotated as an area is eaten down. In early summer one area can be kept ungrazed until a crop of hay has been taken.

Zero-grazing

In a high potential area goats should be kept on a zero-grazing system since land is often in short supply and wandering goats can cause great damage (see Fig. 3.2).

Farmers should confine the goats to a yard and provide hay, farm by-products and cut grass. Under this system there are fewer problems with parasites, including worms. In addition there are no fencing costs and the possibility of young trees and crops being destroyed is eliminated.

Minerals

Milking goats, in particular, need calcium. Feeds rich in calcium such as lucerne and clovers should be given. Other minerals are

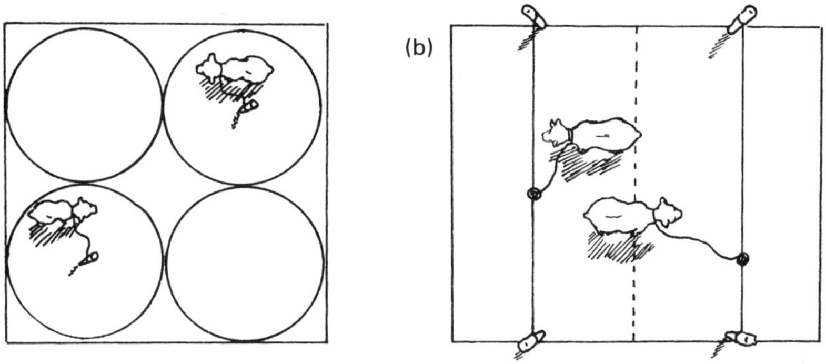

3.2 Tethering goats:
(a) stationary tether, (b) running tether

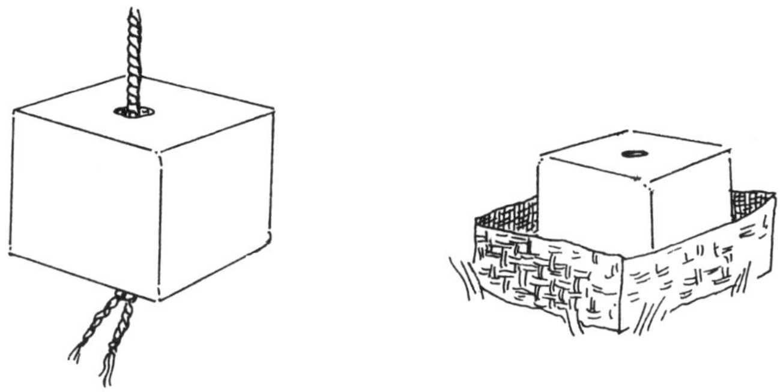

3.3 Methods of providing mineral licks for goats

also essential. A high-yielding goat will draw considerably on her own body minerals so a mineral lick must be available (see Fig. 3.3).

The effect of some mineral deficiencies are:

Cobalt	'wasting' of animal;
Copper	poor reproduction, anaemia;
Calcium	breakdown in high-yielding milkers;
Iodine	still births, thyroid problems, low yields.

Feeding the milking goat

Relative to body weight a good milking goat gives much more milk than a cow. She therefore requires large quantities of nourishing food including minerals.

Some pedigree goats can produce up to 4–5 kg of milk per day but the average is around 1–1.5 kg per day. The quality of feed and grazing will affect the milk yield.

In addition to concentrates (see next section), she requires some 1.5 kg of good hay and some 4 kg of green feed. Heavy milkers require a lot of water and goats also need one teaspoon of salt added for each gallon.

Concentrates

Concentrate can be made up of grains and protein-rich feeds. A concentrate feed should be about 15% protein. Various mixtures can be used depending on the feedstuff available. Some possible mixtures are given below:

2 parts crushed or ground cereal
1 part bran
1 part groundnut meal or soya bean meal
or 1½ parts linseed cake or cottonseed cake
or ground beans

Grain should not be ground up too finely as goats do not like powdery feeds.

Below is a summary of feeding concentrates for various goats:

3–4 week-old goat	a little concentrates
young goats	up to 0.25 kg
pregnant goats	up to 1 kg
male goats	1–1.75 kg
drying off	no concentrates
a few weeks before birth	gradually up to 1 kg
after birth	nothing for 4 days (to prevent a swollen udder) then gradually increase
milking goat	1–1.75 kg depending on milk yield

Milking

Clean milking is just as important in milking goats as in milking cows. For clean milking practices see under 'Milking' (cows), page 26.

The first 1–2 squirts when milking should be discarded and not used as this milk is high in bacteria. Milking should be done twice a day.

The hair on the flanks and around the udder should be trimmed regularly and the goat can also be brushed occasionally.

Milk can smell if goat hair falls into the bucket or if the male goat has been in close contact with the female. Goats can be de-odourized by destroying the scent glands with a red hot disbudding iron before the animal is one month old. The scent glands are just behind the horn base. This should be done by a Veterinarian.

Keep milking utensils clean and wash thoroughly, adding a little disinfectant to the water.

Goats products

Yoghourt, cheese, cream and butter can be made from goats milk. Goats milk is very easily digested and is good for digestive disorders. Some people are allergic to cows milk but can drink goats milk.

Housing

At higher altitudes a warm but well-ventilated house is required. This should consist of a large communal pen and several individual pens. The floor should be made of concrete or rammed earth and should slope towards the door for drainage. There should be feed racks for hay, feed troughs, mineral licks and water troughs. The troughs and water containers should be accessible from outside to save work.

A platform can be made for the goat to stand on when being milked so that she is at a good working level for the milker. The platform should be 70 cm high, 100 cm long and 60 cm wide.

The house can be made of most building materials, but bear in mind the destructiveness of goats. Wood, corrugated iron or brick can be used for the walls, whilst a thatched roof, if out of reach, is best for the roof. The walls between pens need to be some 1.5 m high. Each goat requires some 2.25 m^2 of floorspace (see Fig. 3.5).

3.4 Using a goat milking platform

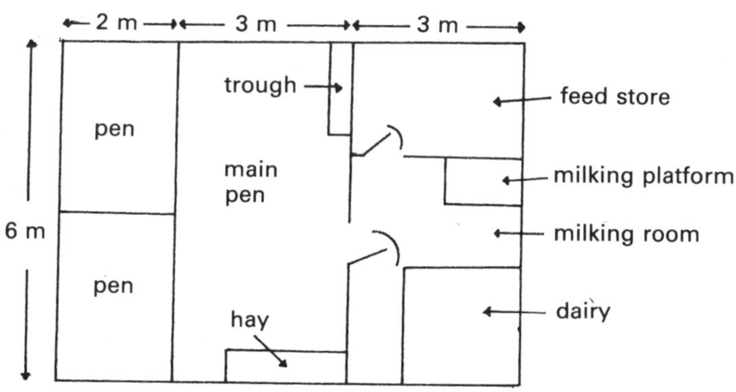

3.5 Plan of goat house for about six goats

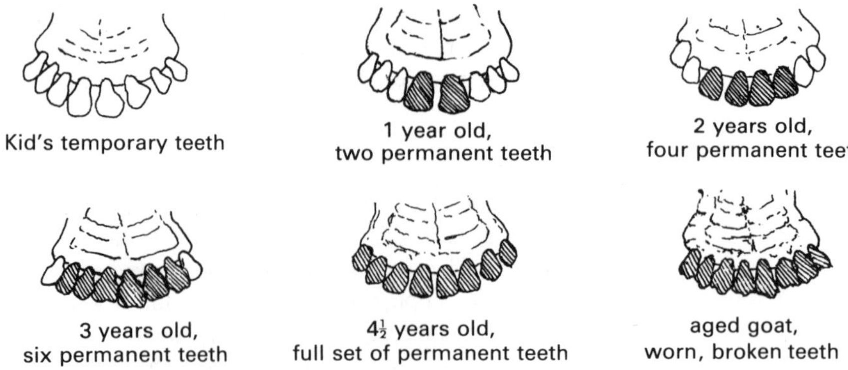

3.6 Estimating the age of a goat by its teeth

Care of hooves

On range land, hooves are worn down to a reasonable level. However, on softer pastures or in a zero-grazing system, hooves grow too long and feet can then become damaged. Every few months the hooves should be inspected and trimmed with a sharp knife. In addition a stone or piece of concrete can be kept in the pen which the goats will often jump on and thus wear down their hooves.

Disbudding

Disbudding prevents horn growth and is a good practice as horns can cause considerable damage. This is done at four days of age.

The hair around the horn bud should be clipped and the bud rubbed with a caustic stick or paste. Alternatively a hot disbudding iron can be used. The iron should be hot enough to scorch wood and should be applied for 15 seconds. If caustic paste or a caustic stick is used then the kid should be restrained for half an hour so that it does not rub the caustic off.

Age of goats

The age of goats can be estimated by examining their teeth on their lower jaw. At 4½ years the goat becomes 'full mouthed' after which the teeth begin to wear down (Fig. 3.6).

Diseases and parasites

For diseases see under 'Sheep' on pages 120–122. In addition the following points are made:

Routine dosing twice a year will control internal worms.

Goats should be deloused every few months with a suitable chemical.

Where goats are kept in low wet areas, they should be dosed against liver fluke once or twice each year.

Health Management

Bloat/Indigestion Caused by eating too much green feed so that the stomach becomes full of gasses. Treat with 100–150 g liquid paraffin and walk the goat about.

Constipation Give water with molasses added to it or drench the goat with 100–150 g of liquid paraffin.

Enterotoxaemia Violent scouring may be caused by enterotoxaemia. This is poisoning from toxins and can be prevented by vaccinating twice a year at regular intervals.

Foot wounds Clean with hydrogen peroxide.

Inflamed eyes If the eye is inflamed and discharging, bathe with warm water to which a little salt has been added.

Mastitis This causes discoloured milk with clots in it. Treat with antibiotic cream squeezing the tube into the teat. Milk the goat last and throw away the milk. Wash hands well. Consult your Veterinarian.

Scouring Signs are loose, liquid droppings. May be due to worms, liver fluke, coccidiosis or to incorrect feeding. Try deworming and give a better diet. If there is no improvement consult your Veterinarian.

Stroke caused by heat Goats, especially white goats, may suffer from this. The symptoms are a high temperature and staggering. Provide adequate shade.

Poisons If the goat appears to have eaten a poisonous substance, drench with 100–150 g liquid paraffin.

4 Pigs

Introduction

Intensive pig production entails considerable capital outlay and specialised management knowledge. Any one considering pig production on a large scale should investigate the marketing and processing facilities available and obtain expert advice. Many farmers have small herds of 5–10 breeding sows. Unfortunately, many pig farmers keep their pigs in rather poor conditions and so are not able to make as much profit as they should. Pigs are particularly affected by dirty, drafty housing and quickly become sick. They are also affected by poor feeding particularly as they are kept in pens and so cannot find or root for extra or more suitable food.

Breeds of pig

Some of the main breeds are described below and illustrated in Fig. 4.1.

Duroc

This has drooping ears and is light to dark brown in colour; it does not produce a very good carcass but is an efficient feed converter. The Duroc is a hardy animal which survives well in tropical climates; it is generally used for cross-breeding.

Large white

This pig has erect ears, is white coloured and very prolific. The breed can be used for both pork and bacon production. It is a fairly hardy animal but will suffer from sunburn if it is not kept in a building out of the sun.

Tamworth

This hardy animal of a reddish colour and with erect ears is an efficient converter of food. Its long body makes it excellent for bacon production.

Landrace

With drooping ears and white in colour, this pig is good for bacon production but requires a high level of management.

4.1 Pig breeds

Wessex Saddleback

A black pig with a white band over the shoulder and white front legs, it is very hardy, but rather too short and too fat for bacon production. It is generally used for crossing with other pigs.

Hampshire

This pig is very prolific. It has more meat than the Large White and Landrace but is probably best kept for cross-breeding.

Local breeds

There are many local breeds of pigs in different countries. These are usually fairly resistant to diseases and heat stress and are sometimes used for cross-breeding. Work is being undertaken with some breeds to improve their productivity through breeding programmes.

Management

Housing

The main types of houses are described below.

Danish pig house

A fattening house with a central feeding passage along the walls.

Traditional pig house

For a small farmer a series of simple pens with an outside run attached may be sufficient (see Fig. 4.2). Pigs require warm dry

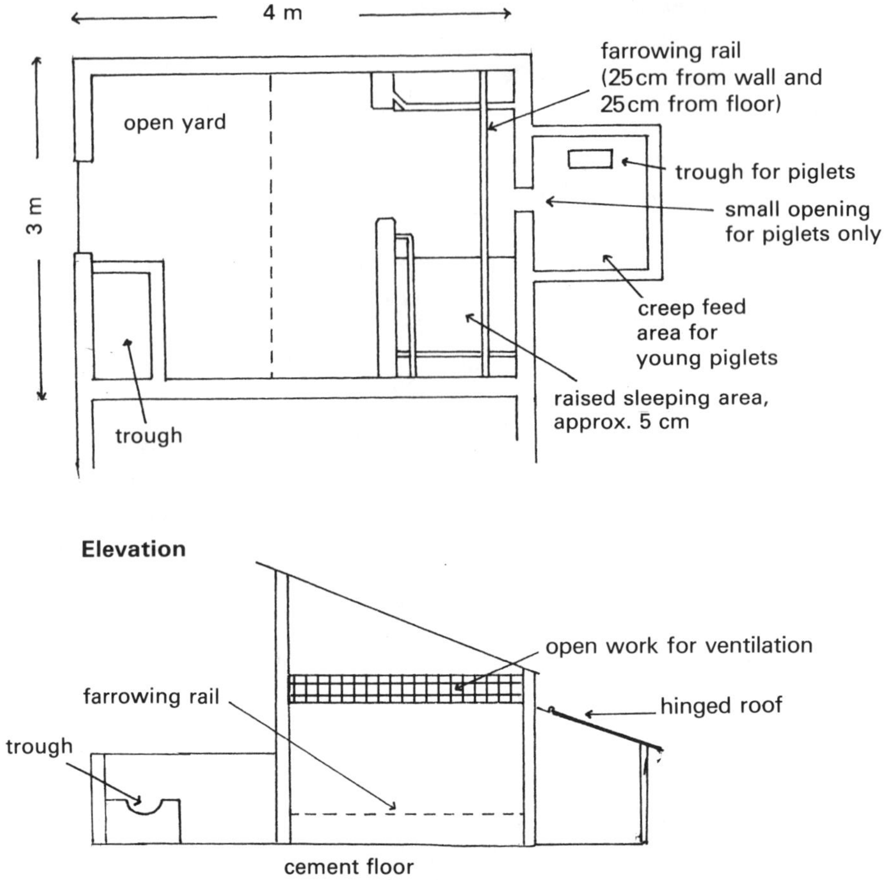

4.2 Housing for sow and her litter

housing, free from draughts, as they easily get pneumonia. The floor should be well drained, or concreted, and the unit not too near the dwelling house because of smells and flies.

If it becomes cold, a false ceiling can be made in the pig house from chicken wire and straw. This should be about 1.5 metres from the ground. Pigs are liable to get pneumonia if they get cold.

Field housing

This can be a simple shelter made of wood, corrugated iron, etc. The shelter is in a field or wood and the pigs have free range. Generally it is not advisable to keep pigs outside in the tropics as they are then subject to African swine fever and kidney worm.

Feeding

The three pig feeds usually available commercially are: creep feed (some 16 percent protein); sow and weaner meal (14 percent); and pig finishing meal (12 percent) for fattening up pigs for pork and bacon.

Minerals are very important and rations must contain some chalk (calcium), bone-meal (phosphorus) and common salt (sodium chloride).

A sudden change in feedstuffs upsets pigs and prevents them from gaining weight. Changes in feedstuffs should be made as gradually as possible, over a period of a week or more.

Table 4.1 Approximate daily food allowances for different classes of pigs

Boars	3 kg sow and weaner meal (less if getting too fat)
Dry sows	2.0–2.5 kg
Farrowed sow	1 kg plus 0.5 kg for each piglet
Fattening pigs	
8 weeks	1 kg
10 weeks	1.25 kg
12 weeks	1.5 kg
14 weeks	1.75 kg
16 weeks	2 kg pig finishing meal (change gradually)
18 weeks	2.5 kg
20 weeks	2.5 kg
22 weeks	2.75 kg
24 weeks	3 kg
26 weeks	3 kg
28 weeks	3 kg

Pig meal can be purchased from millers. Sometimes, however, commercially-made meal is difficult to obtain. A farmer may then have to mix up his own meal from whatever he can obtain. A homemade feed for pigs should contain both energy foods and protein foods.

Energy foods are maize, wheat, rice, barley, sorghum, oats, cassava, sweet potato, Irish potato, fruit pulp and cereal brans. Protein foods are oil-seed meal (for example, cotton-seed, soybean, linseed, groundnut, coconut or sesame-seed meals), beans, peas, blood-meal, meat-meal, fish-meal or skim milk. In the case of both energy and protein foodstuffs, it is best to mix a number of different types and not to feed one type alone. Some 10%–20% of the mixture should be protein food. Fattening pigs require least protein whilst nursing sows and piglets on creep feed require most. The mixture should also include 0.5% of each of the following: bone-meal, salt, a vitamin mixture and calcium (chalk).

Where possible the farmer should obtain professional advice on pig feeding as it is complex and dependent on available feedstuffs.

Water requirements

Some farmers provide water at all times, whilst others prefer to give regulated amounts, although this should be increased in hot weather.

Table 4.2 Water requirements of pigs

Type of pig	*Water requirement for each kilo of feed* (litres)
Fattening pigs	2.5
Lactating sows	5
Dry sows	3
Piglets	water always available

Breeding

There are few pedigree pig breeders in the tropics as yet. Most farmers raise non-pedigree stock and sell weaners, finished porkers and baconers.

Selection of breeding gilts

Gilts should have at least 12 teats so that there will be enough teats even for a large litter. Select gilts from sows that wean 9–10 piglets per litter, and who are known to be good mothers.

Select fast-growing pigs as these pigs are likely to consume less feed per unit of liveweight gain. Selected pigs should have good strong legs, adequate length, strong topline, well-developed hams and comparatively light heads.

Selection of the boar

The boar contributes half the quality of the herd and thus it is extremely important to select a good one. The main points to look for are:

(a) he should come from a fertile mother and a father whose other offspring are satisfactory;
(b) he should have sound feet with good, full hams;
(c) a uniform curve of the back and a good length;
(d) at least 12 nicely placed teats, so as to pass on this characteristic to his female offspring.

Management of breeding stock

Boars

A boar should not start serving until he is over eight months of age; and during the first two months of service, only twice per week.

Subsequently, he can be used for up to six services per week. One boar can look after 15 sows. The first service after a rest period should not be counted as the semen may not be fertile.

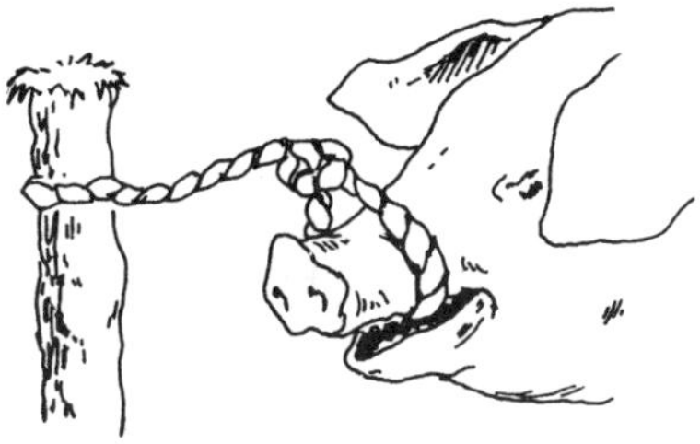

Tethering

Guiding pig forward with solid boards

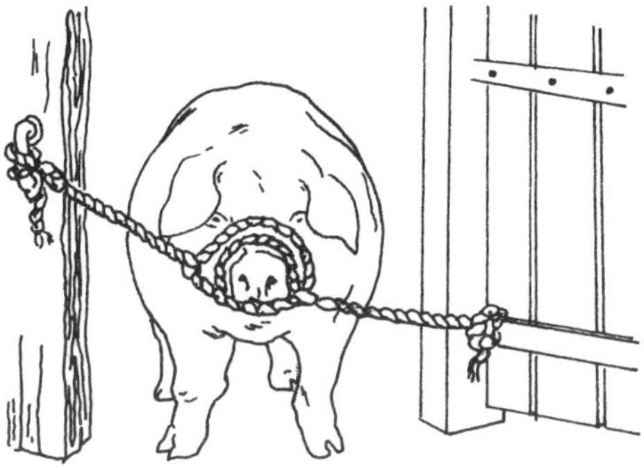

Securing for veterinary treatment

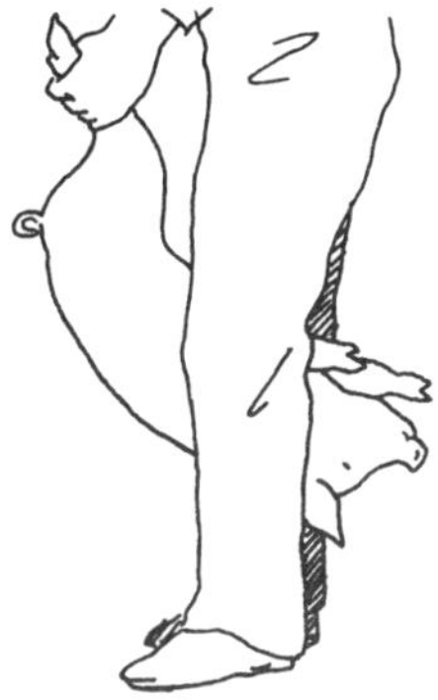

Holding for castration

4.3 Handling pigs

Considerable exercise is necessary to prevent the development of leg weaknesses. It may be necessary to trim the boar's feet regularly. Boars should be washed with soap and water every four months and sprayed for lice and mange. The pen walls should be whitewashed with a wash containing Gamma BHC at the same time.

Gilts/sows

Gilts and sows will tend to get too fat if they are not allowed enough exercise. A fat sow takes longer to come on heat and is more likely to crush her young piglets. Sexual maturity occurs as early as four or five months, but the first service should not be until 8 months, when the weight should be 100–130 kg. A sow has a productive life of four to five years.

Breeding cycle

The normal heat period lasts for three to five days; signs of heat are:

1st stage:
 general restlessness,
 vulva turns red and swollen,
 white mucous discharge;

2nd stage:

real heat lasts for 40–60 hours,
vulva less red and swollen, slimy mucous discharge,
tendency to mount and be mounted,
the sow or gilt will stand still when pressure is applied to her back (can accept a man's weight sitting on her),
she is ready for service and this is the right stage to send her to the boar;

3rd stage:

after the heat, the sow will not stand still when pressure is applied to her back,
the swelling of the vulva disappears.

Recommended practices

Usually only a few of the signs will be seen and therefore it is best to put the sow with the boar for a short period every day when the heat is expected. Always take the sow to the boar. This is less upsetting for him. It is best to put them together just before feeding. Let the boar serve twice, with an interval of about 12 hours between services. If the sow does not conceive, she will return on heat in about three weeks' time. The gestation period is 114 days (3 months, 3 weeks, 3 days). Pregnant sows should not be too crowded in their pens as this can cause abortion.

A week before service, give the sow/gilt 1 kg of feed extra per day (standard is 2.5 kg/day). Continue this for one week after service. During the last month of pregnancy give 0.5–1 kg extra feed per day, but decrease this gradually one week before farrowing, and provide plenty of water to help prevent congested gut during farrowing.

Preparation for farrowing

(a) The farrowing pen should be dry and free from draughts.
(b) Clean the farrowing pen thoroughly and scrub with strong tar disinfectant. If possible, apply a whitewash mixed with Gamma BHC or similar insecticide to the pen walls to control mange and lice, at least ten days before the expected farrowing.
(c) Deworm the sow two weeks before the expected farrowing.
(d) Move the sow to the farrowing pen four to seven days before the expected farrowing date, so she can familiarise herself with the new surroundings. She should be washed with soapy water and weak disinfectant to remove dirt and parasite eggs.
(e) Provide plenty of straw, chopped into short lengths of about 10 cm to prevent the piglets being trapped in longer straw and being crushed by their mother. The first three days are the most dangerous for the piglets.
(f) One day before farrowing the sow becomes restless and later milk can be squeezed out of the teats. She will begin to make a nest with straw.
(g) Use a well designed farrowing pen to help prevent the sow

from accidentally crushing the piglets. The farrowing pen has a rail around the wall, behind which the young piglets can escape when the mother lies down. Some farmers now use a farrowing crate, which minimises piglet losses through crushing.

Care of piglets

Heating for piglets

In cold weather, a small area can be heated with an infra-red lamp, or a 50 watt electric light bulb hung 30–40 cm above the piglets. This keeps the young pigs warm, helps prevent pneumonia and crushing, as the piglets tend to stay under the lamp when not feeding. In addition warm bedding, chopped straw or sawdust, etc., should be put on the floor.

Creep feeding

Young piglets from 10 days onwards should have a high-protein feed available to them. This has to be fed in a small creep or area where the mother cannot eat the feed. The feed conversion rate of young piglets is very high and thus creep feeding is particularly economic.

Creep feeding is very important as the piglets become used to feeding on meal at an early age; the sow's milk yield also begins to decrease just as the piglets require more feed.

If piglets have scours then a commercial antibiotic can be added to their drinking water.

A farmer who has more than a few sows should earmark the piglets, weigh them and record the weight.

The canine and premolar teeth should be cut to prevent damage to the sow's teats, as they may be 'needle' sharp. This can be done with a pair of teeth cutters, within 24 hours of birth.

Heat stress

In very hot areas a concrete 'wallow' (small pond) can be constructed and the boars and pregnant sows particuarly, be allowed to wallow 3–4 times a day. It must be kept clean because of disease and parasite infection risks. Alternatively the pigs can be sprinkled several times a day with water, to reduce their body temperature.

Provision of iron

If the piglets are not on pasture, iron should be provided in the form of an injection of 2 ml of iron-dextran, or a paste painted on the teats or fed into the mouth of piglets, or by providing clean soil. A square of grass and soil can be put in the pen each day. Injections should be given at one day old. Piglets suffering from lack of iron become anaemic; they have white diarrhoea and frequently die.

Castration

Male piglets should be castrated at two to three weeks of age unless wanted for breeding. This should only be done by an experienced stockman and therefore details are not given here.

Weaning

Weaning should take place when the piglets are eight weeks old, and weigh about 13–16 kg. Some commercial farmers wean at 5 weeks of age but this is only possible when there is a good 'pre-starter' meal available. Remove the sow but leave the piglets in the pen as this lessens the stress to the young piglets.

A few days after weaning, gradually change the feed from creep meal or pellets to sow and weaner meal. After one week, deworm the piglets and move them to the fattening pen. The sow will come on heat again two to seven days after weaning.

Tail biting

Some farmers cut the piglets' tails during the first week. This prevents tail chewing which can set up infections, but it is not usually necessary. A piece of chain can be hung down from the ceiling for the piglets to chew, instead of the other piglets' tails.

Culling

Sows that are difficult to get into pig (pregnant) or which only manage to rear small litters should be sold off; also boars which are infertile or moderately infertile.

Records

Records should be kept so that only the most efficient sows and their offspring are kept as breeding stock. Useful records are:

(a) Litter records: birth weight and weaning weight; 1.5 kg and 18 kg are good standards respectively.
(b) Dams records: number of piglets weaned per year; 18 piglets is a good standard.
(c) Marketing: age and weight; 100 kg at six months old is a good standard.
(d) Food conversion rate: pigs which gain more weight from a given amount of food are good converters. A satisfactory conversion ratio would be 1 kg liveweight gain for 3.5 kg of feed for pigs which are being fattened.

Marketing

The farmer should weigh his pigs regularly to check on their performance and he will thus be able to take them to market at the best time. For pork, the pig should weigh 45–55 kg liveweight, and for bacon 80–110 kg liveweight, before slaughter.

Commercial farmers weigh pigs every week as a check on satisfactory management.

Parasites and diseases

Parasites

Roundworms

Roundworms are a major problem, and routine dosing is recommended: boars, every six months; sows, two weeks before farrowing, and after weaning; piglets one week after weaning; and fattening stock four months.

Lice, fleas, mange

Scrub sows with Gamma BHC insecticide four days before farrowing. Clean and scrub out the farrowing pens before use.

Pig diseases

African swine fever (Warthog disease)

A highly contagious viral disease of pigs. The symptoms are similar to those of hog cholera, but pigs resistant to hog cholera will succumb to African swine fever. The infection in domestic pigs often follows contact with warthogs or bush pigs, which serve as carriers.

Symptoms High fever for four days (105°F). As temperature declines, affected pigs lose their appetites, become depressed, weak and uncoordinated. Coughing is evident in some animals, while others develop a bloody diarrhoea. Death occurs within 48 hours following these symptoms, and six to seven days after the initial fever. Mortality in domestic pigs is nearly 100 per cent.

Treatment None. Infected herds must be slaughtered; the housing should be disinfected thoroughly with a 4 per cent solution of caustic soda and should be kept empty of pigs for two months.

Control Impose quarantine measures to delay spread of the disease. It is a notifiable disease in many countries (p. 52).

Swine dysentery

A disease of swine of all ages, also known as bloody scours. The symptoms are somewhat similar to hog cholera.

Symptoms Dysentry affecting all ages but most severe in younger pigs; the diarrhoea has blood and mucous in it. Animals may show a fever.

Treatment Use of antibiotics will reduce death losses.

Control Isolation and quarantine; newly-purchased pigs should be quarantined for two to three weeks.

Swine erysipelas

An infectious disease affectly mostly young pigs. Stress, for example a sudden change in diet, may bring on the disease.

Symptoms These may vary and include sudden death, or sleepiness, loss of appetite and fever. Skin lesions also occur, in which case the disease is referred to as diamond skin disease. In the chronic form, lameness due to swollen joints may occur.

Treatment Penicillin and serum combined, as prescribed by Veterinarian.

Control Vaccination. The disease is classed as notifiable.

Hypoglycemia

Also known as baby pig disease or three-day pig disease as it occurs in the first few days of life.

Symptoms Shivering, hairs standing up, squealing, weakness, fall in temperature, unco-ordinated eye-balls and twisted necks. Without treatment, death usually occurs in 24–48 hours.

Treatment Provide warmth; feed glucose by injection.

Control Select breeding stock of quiet temperament and high milk yields. Provide warmth for pigs farrowing in cooler seasons.

Transmissible gastroenteritis

A rapidly spreading viral disease of swine.

Symptoms Scouring; piglets discharge yellow-greenish or white faeces. Mortality is up to 75 per cent in piglets less than one week old. In older animals, watery diarrhoea, decreased appetite, loss of weight and sometimes vomiting occur. Recovery is usually quick.

Treatment No specific treatment. Give piglets a milk replacer or cow's milk if the sow's milk supply has decreased.

Control Strict hygiene and prevention of visitors to the buildings must be practised. House farrowing sows as far apart as possible. Dispose of dead animals by deep burial promptly, thus avoiding scavenging. Do not introduce new stock during the farrowing season.

Swine vesicular disease

A highly infectious virus disease which may be confused with foot and mouth disease. The source of infection is mostly raw garbage containing infected pork scrap.

Symptoms Lameness, blisters on or about the feet, snout, lips, gums and tongue. Animals may show a fever (104°–170° F) and loss of appetite. Young pigs are more severely affected. Abortions of near-term sows and other breeding difficulties occur. Secondary bacterial infection may follow.
Treatment None.
Control Isolation and quarantine. Garbage fed to swine must be cooked.

Leptospirosis

See 'Cattle diseases', p. 61.

Tuberculosis

Important where bovine or avian tuberculosis has been diagnosed. See 'Cattle diseases', p. 57.

5 Poultry

Commercial poultry production

In many countries the main poultry production is undertaken by large commercial companies. However, this section is concerned with the small commercial or home producer who keeps a small flock. Eggs are an excellent protein source and to produce even a few on a regular basis can improve the diet of a farm family.

Main breeds

Some of the main chicken breeds are listed below.

White leghorn

A light breed very suitable for commercial egg production; can be nervous and flighty and requires good management.

Rhode Island Red

Dual purpose; easily managed, fair layer with good carcass, although yellow-skinned.

Light Sussex

Dual purpose and easily managed.

Black Australorp

A heavy bird, moderate layer.

First crosses and hybrids

Specialised breeders sell chickens which are first crosses and multiple crosses (hybrids). These gain weight more quickly and lay more eggs than pure breeds and are therefore generally used by poultry producers.

Type of birds

These are classified according to the use of the bird.

Hybrid broiler These are used for commercial meat production only. They have a good food/meat conversion ratio. A mature female should weigh over 2.75 kg. Young broilers fatten very quickly and are ready for marketing at twelve weeks of age.

Dual purpose birds These are relatively disease-resistant, have mild temperaments and rarely exhibit cannibalism. They also give good carcasses on slaughter. Their egg production is only fair

5.1 Poultry breeds

and they often have a tendency to go broody. Usually recommended for new poultry farmers. Mature females weight around 2.25 kg.

Lightweight birds bred for egg production These birds have an excellent food conversion rate and rarely go broody. Egg production can be very good. Mature females weigh under 2.25 kg. They give small carcasses on slaughter and are also inclined to be nervous and exhibit a high degree of cannibalism. These birds require good feeding and management. They are recommended for experienced poultry keepers only.

Management systems

There are several systems used in poultry management:

Free range

Birds are allowed to range over a pasture. The houses are placed in the field. The system takes a lot of labour; egg production is not very good and the birds are not secure from theft.

Raised floor system

The poultry shed has a floor made of chicken wire and raised 1–2 feet so that the manure falls through. It is fairly expensive because of the floor and the regular cleaning necessary.

Battery

Hens are kept in cages all the time, usually two to three hens in each cage. The cages are inside a building. The costs are very high and this system is only suitable for commercial producers following very skilled management practices.

Deep litter

The birds are confined in a house with chopped straw or wood shavings spread on the floor – about 10 cm deep. This system can be used by large and medium-sized producers. With careful management this system will give good results and is recommended as suitable for most farmers. The management system for deep litter production is described in the following pages.

The management of litter

The litter on the floor must be kept dry or disease may spread through the flock. If the litter becomes damp and smells of ammonia, then it can be replaced, or lime can be sprinkled on the litter and the litter turned with a spade. Additional litter can be added from time to time.

Water containers should not be overfilled or liable to leak as wet places in the litter will become breeding places for worms and disease.

To prevent boredom which may lead to cannibalism, and to encourage exercise in the deep litter house sprinkle some grain in the litter to encourage the birds to scratch for it; also hang up some greenstuff so that the hens have to jump to reach it.

Buildings

These should be rainproof, and have smooth surfaces, so that mites and other pests cannot hide. Almost any materials can be used for building walls. However, smooth plastered walls are best as insect pests cannot hide so easily. Good ventilation is necessary and in hotter areas at least two walls can be of chicken wire only. Good ventilation helps to prevent damp and respiratory diseases.

The floor should preferably be of concrete, although wood, earth or sand can be used, but these are difficult to keep clean and harbour disease organisms and pests.

Ideally houses should be ratproof, as rats consume food and may damage young chickens.

For litter use 75 per cent wood shaving and 25 per cent sawdust as this makes a good combination. However, groundnut shells or chopped straw or crushed maize cobs can be used.

Care of young

Some poultry specialists sell eight-week old birds to poultry farmers. This saves losses and the problems of managing a brooder. The pullets are vaccinated against Newcastle and

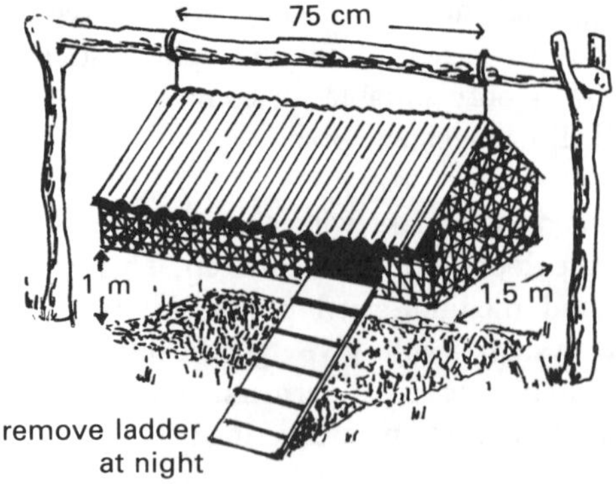

Suspended poultry house for about 10 birds

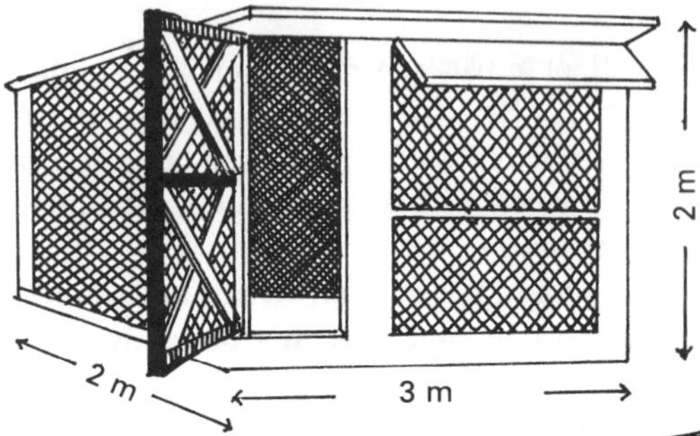

A simple chicken house for hot areas, for about 30 birds. Birds outside in day time.

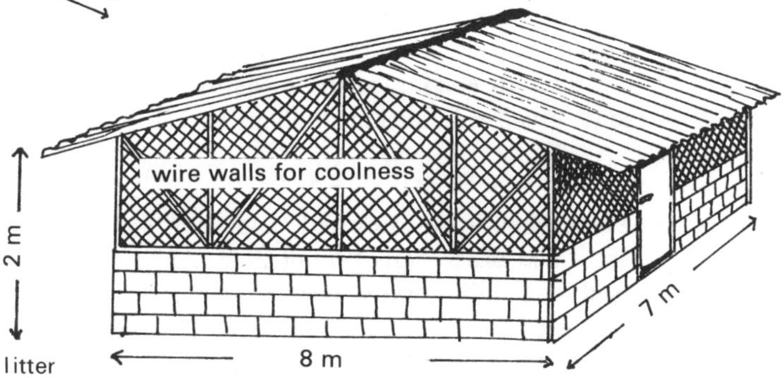

Deep litter house for 200 layers, warm climate

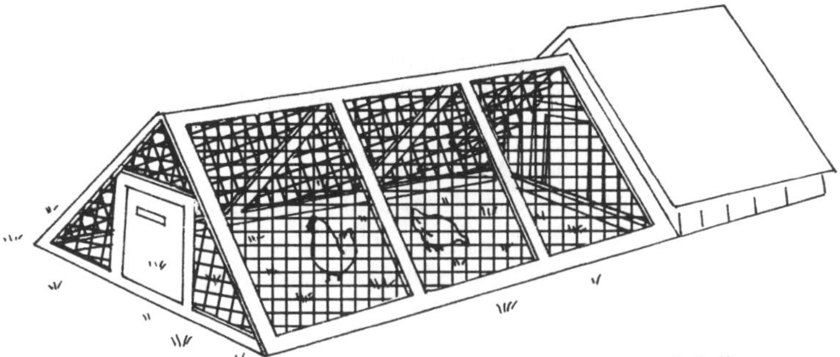

Fold unit, should be moved daily

5.2 Poultry housing

Marek's diseases before sale. Many poultry farmers purchase day-old chickens from commercial hatcheries. Before the day-old chickens arrive on the farm, the farmer must have the following ready.

(a) A brooder house. This is usually a small shed, which is safe against bad weather and predators. Here the day-old chickens live until they are 3–5 weeks old. It should be washed with soapy water and disinfectant several days before use.

(b) A heating unit. If this is a hurricane lamp then wire meshing should be put around it to protect the chickens from burning themselves.

(c) Litter covering the floor of the brooder with a 1–2 cm layer, this can be preferably wood shavings, but ordinary chopped straw or hay can be used. The floor can be covered with plastic or paper instead of litter to prevent the chickens from mistaking any litter for food.

Up to four weeks old, an area of 1 square metre will accommodate 20 chickens. At two months, however, this space will only be adequate for 10 chickens.

The farmer must ensure that the brooding equipment is functioning properly, i.e. that he can control the temperature and ventilation.

Paraffin lamps

Paraffin lamps need careful handling. They should never be more than three-quarters filled, because of the likely build-up of vapour and danger of fire. Lamps should be lit the day before so that any adjustments can be made. Fire is a big danger and every care must be taken with lamps. The brooder should be away from other buildings.

Warm-floor brooder

If the brooder is a warm-floor brooder (being heated from the underneath) then the chickens will be on a wire mesh floor. This should be covered with a sack for the first week as their feet are small. Later the wire mesh floor helps to prevent coccidiosis, as the chickens cannot eat their droppings.

Infra-red lamps

If a farmer has electricity he can use an infra-red lamp. Infra-red lamps are hung from the ceiling and the height needs adjusting as the chicks grow. A fairly large circle made of roofing felt or timber can be made around the lamp. Chicks should be able to move into the warm area and out again if they wish.

Overcrowding

The brooders must not be overcrowded as otherwise the chicks may get crushed. Corners should be rounded for the same reason. Use a barrier to keep chicks near the lamp for the first 10 days. Then it can be removed.

Food and water

Chicks should not be allowed to foul the troughs, so after two weeks the troughs can be hung from the ceiling. Wire mesh should be fixed over the water containers to prevent them from getting wet. Provide fresh clean water in a number of drinkers from the time the chicks arrive and are placed in the brooder.

Temperature control

Temperatures in the tropics can vary considerably between day and night. It is essential to have a thermometer and to read it several times each day and evening. Chicks that are too cold will suffer from chilling and may die. Careful temperature control is therefore essential.

The temperature requirement for day-old chicks is 35°C and by the end of the first week is around 27°C. Reduce the temperature gradually as the chicks mature, especially in the middle of the day, but do not let it go below 18°C. If the weather is cold keep the temperature up particularly in the evenings and at night. If the chicks huddle under the heat they are not warm enough. If they go as far from the heat as possible then they are too warm.

Heat is necessary for the first five weeks. For the next three weeks the chicks should remain in the brooder house without heat. At eight weeks the chicks are moved to the rearing house.

Feeding

For the first eight weeks, feed commercial chick mash. At eight weeks old gradually introduce growers mash or pellets.

By week 20 (or when the first egg is laid) gradually introduce layers mash. Also give limestone, oyster shell or marble chipping to improve eggshells. Without this shells may become very thin. This can be provided in a small tin or box.

A tin full of grit or coarse sand should be provided. This grit is retained by the chicken in its gizzard in order to grind up food.

Vaccinations

Vaccinate against fowl pox at five weeks and against Newcastle disease at eight weeks, before the birds are moved to the rearing house. Vaccinate against fowl typhoid at about 16 weeks.

Rearing pullets

At eight weeks the pullets are moved to a rearing house where they will remain until moved to the laying house at around 20 weeks of age.

The house and equipment should be washed and disinfected several days before. Clean litter should be spread on the floor. Roosts should be provided.

If the birds are too crowded they will not do well. Pullets require a minimum of 0.5 square metres each and 10 cm of trough space. The corners of the house can be filled with litter to prevent crowding into corners on cold nights and causing suffocation.

Pullets should not be mixed with older birds because of the danger of diseases being passed on.

Land which has been used for a time by poultry will be a possible source of internal worms and coccidiosis and a dressing of lime will be very useful to disinfect the ground.

Feeding layers

A layer will need 120 g of food per day, preferably a commercial foodstuff mixture. This can be a mash or pellets. Trough space should be 13 cm per bird and the troughs should be cleaned weekly to remove stale food.

The food troughs and waterers should be evenly distributed in the laying house so that birds do not walk more than 10 metres from food to water. There must always be water available otherwise the hens may moult. Provide grit to enable the birds to digest food.

Laying birds require calcium, and marble chippings should therefore be provided as these are a cheap source of calcium. Fresh green feed can also be provided, up to some 3 kg per 100 birds each day.

Laying boxes

If daylight hours are short, production can be increased by keeping lights on for several hours, giving the birds longer feeding time; 13–14 hours of light each day is best.

The laying boxes (Fig. 5.3) must be kept dark to prevent egg-eating, and the eggs should be collected three times each day. The boxes should be lined with clean straw and dusted monthly with pyrethrum powder against mites. Allow 20 nests for every 100 hens.

Broody hens

A broody hen is one that has stopped laying and spends her time sitting on eggs in the nesting boxes. Remove broody hens from boxes to prevent damage to eggs. Put the broody hens in small well-ventilated cages for several days to discourage broodiness. Also watch out for egg-eating hens that peck the eggs.

Floor space

The minimum floor space required for poultry is given in Table 5.1.

Table 5.1 Required floor space for poultry

Age in weeks	*Birds per square metre*
0–4 (brooder house)	20
4–8 (brooder house)	10
8–20 (rearing house)	5
20 or more (laying house)	4

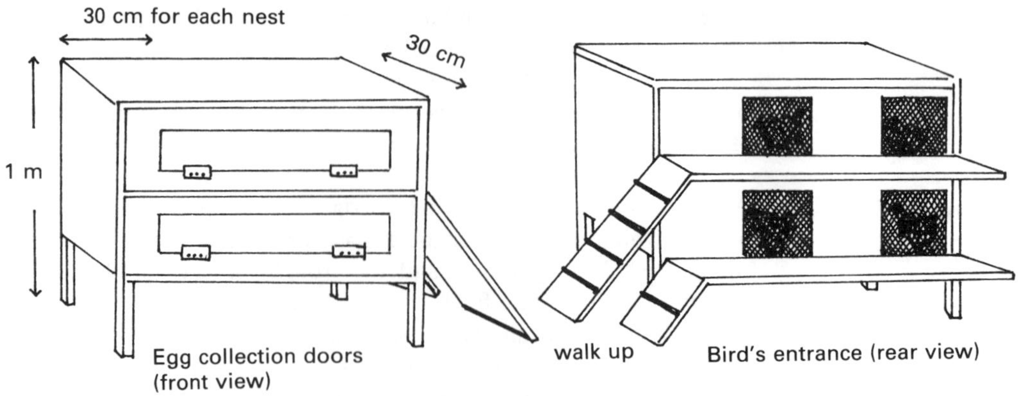

5.3 Nesting box for poultry

Culling

Start culling if the egg production drops below 40 per cent of birds laying each day. A laying hen can be recognised by her bright red comb and wattles (a hen which has laid a number of eggs may have a rather faded comb and wattles). She should have bright and alert eyes, and measure at least two fingers width between pelvic bones and the end of the breast bone. Her feathers may be rather ragged. Birds should be examined regularly and if they do not exhibit any of these signs they should be culled (sold off or eaten). Birds are normally kept for two or three years in the laying flock.

Management problems

Egg eating by hens This can be cured by debeaking (cutting off a small part of the upper beak) or by culling out the egg-eating birds. Nesting boxes should be sufficient in number, and dark.

Early moulting This may be due to poor management, such as frightening the birds, or lack of water for a period. Try to feed, water and collect eggs at the same time each day and with the minimum of movement and noise.

Cannibalism This occurs when there is too much light or too crowded conditions, or when a wounded bird is attacked by the others.

Signs of sickness

Birds that are sick may be identified early by their general dullness and lack of activity. The tail feathers may be dirty and the bird may remain by itself. The poultry keeper should observe his flock every day and remove any birds that look sick. They should be kept in small individual pens away from other birds. This action may save disease spreading through the flock.

Disease control

Some causes of diseases are as follows:
(a) poor ventilation which causes respiratory diseases;
(b) poor quality feed or feed not mixed in the correct proportions; dirty or impure water, which cause diseases;
(c) putting young birds in with older worm-infected birds;
(d) lack of cleanliness in the poultry house, and dirty house walls and perches providing a home for pests; houses and perches should be washed each time the poultry house is free of birds (once or twice a year);
(e) overcrowding causing stress and weakening of the birds;
(f) dead birds must be removed and sick ones isolated;
(g) allowing visitors into the poultry house who may carry disease on their shoes;
(h) lack of a foot-bath with disinfectant at the entrance to the house;
(i) failure to vaccinate the birds against Newcastle disease.
Thus many diseases can be prevented by good hygiene practices.

Marketing of eggs

Collect eggs at least twice a day to prevent breakages and dirty eggs. Provide soft, clean litter in the nests and change it often to keep eggs clean.

Before packing, remove broken, deformed, or badly soiled eggs and also eggs with blood spots. These latter can be seen through a strong light; this examination is called 'candling'. Market the eggs at least twice a week, so as to give your customers fresh eggs. Store eggs with small end down on egg trays so that the air space is upwards. Wipe dirty eggs with a damp sponge.

Records

Poultry keepers should keep records so as to be able to monitor their enterprise.

Non-commercial poultry keeping

Many farmers keep poultry in a small way just for their own use. Their expenditure then needs to be kept down to the minimum necessary. However if no expenditure is made and poultry are not fed and housed properly then results will be very poor.

Types of birds

Medium sized, dual purpose, hardy birds should be kept. If fed properly these can lay 100–150 eggs per year and after, say, two to three years of production, can still provide good carcass weight. Under good management it may be profitable to purchase hybrid chickens.

Management system

On a free range system the birds will remain healthy and find some of their own food in the form of grass and insects. They should not be confined in a small permanent pen where they are liable to pick up diseases and parasites. The house can be plastered wattle-and-daub construction with a thatched roof. It can be raised off the ground on legs to discourage predators, or even hung in a tree.

If a deep litter house system is used with no free range, then a commercial poultry feed and litter will be required and costs increased.

Feeding

On free range, birds can be fed a homemade ration of cereal and beans. This could be made up of one part ground beans to two parts ground maize or wheat.

Birds can be fed kitchen waste and green feed, and dried blood can be obtained from a butcher. They will find their grit needs for themselves on free range. However, laying birds still require a calcium intake and this can be provided by a supply of crushed eggshells.

Water should be available at all times otherwise egg production will suffer. Keep the water trough in the shade. Feed and water troughs should be cleaned regularly. Birds like to perch on these containers and they foul the feed or water. A bar which rolls easily on top of the feed trough will discourage this (Fig. 5.4).

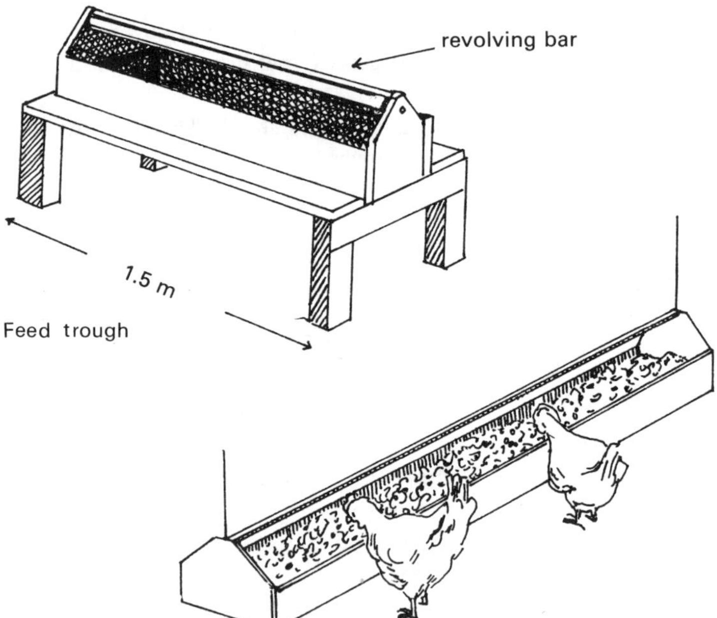

5.4 Feed trough and suspended feed tray for poultry

Breeding

A cock is not necessary for egg production, only if the farmer wishes to raise his own chicks.

One cock is needed for 10 hens. Some hens will go 'broody' and one can be put in a broody house to sit 8–12 eggs depending how big she is. She requires food and water near and there should be a nest of dry grass in the house. The hen should not be disturbed. The eggs will hatch on the 21st day. Alternatively, chicks can be purchased from a hatchery and put under a broody hen. The chicks should only be introduced at night so that the hen does not reject the chicks.

Rearing

Finely ground food and water in shallow troughs should be available for the chicks. The broody hen will look after the chicks. However the house floor must be kept clean to prevent diseases and a wooden or wire fence may be necessary for protection against predators.

After three weeks it may be necessary to move the family to a larger house, and after 10 weeks to provide perches. Surplus cocks can be sold or eaten at about five to six months of age.

Disease control

Have the chicks vaccinated against Newcastle and fowl pox diseases before they are two months old. Newcastle vaccine is required every six months. Vaccinate against fowl typhoid at four months.

Nesting boxes

Dark and secluded nesting boxes should be provided. One nest box is needed for every five laying hens.

Common diseases and their control

The poultry farmer should identify signs of disease by observing:
(a) drop in egg production or growth rate;
(b) drop in water consumption;
(c) general appearance of the birds looking sickly and listless.

Serious diseases

The following diseases of poultry are very serious. In many countries an outbreak or suspected outbreak of any one of them must be reported to the veterinary department, and may result in destruction of the whole flock.

Fowl pox

Also referred to as avian diptheria, canker or sore head and is caused by a virus. Transmitted by contact with infected birds, flies or mosquitoes or free-flying birds. (See Fig. 5.5.)

Symptoms Loss of appetite, dullness, infertility. Wet pox:

5.5 Poultry diseases

discharge from eyes and nostrils, cheesy membrane in mouth; birds suffocate unless membrane is removed. Dry pox: unfeathered parts of skin, comb, wattles and legs show blisters and wartlike scabs.

Treatment Vaccinate unaffected birds; add streptomycin sulphate to feed; add antibiotic/vitamin mixture to drinking water. Infected birds may have to be destroyed.

Prevention Vaccinate birds in areas where the disease is prevalent.

Newcastle disease

This disease is caused by a virus and is very serious. It is spread rapidly by contact from one bird to another and by use of dirty equipment. The chicks gasp and cough leading to a lack of co-ordination, paralysis and coma. Up to 90 per cent of the chicks may die. Symptoms in adults are coughing, occasional paralysis, soft-shelled eggs; egg production may cease altogether. (See Fig. 5.5).

Treatment No drugs are effective against Newcastle virus. Prevent secondary complications by stimulating appetite with wet mashes containing antibiotics or antibiotic/vitamin mixtures in the drinking water.

Prevention Vaccination. Avoid introduction by refusing visitors access to flock and segregate birds of different ages.

Ornithosis (psittacosis)

Caused by a micro-organism, this is a serious disease which may be transmitted to humans causing high fever, difficulty in breathing, pneumonia and sometimes death.

Symptoms Loss of weight, difficulty in breathing, mucous discharge from the nostrils. Some birds may show very little signs of illness.

Treatment Addition of chlortetracycline to feed.

Prevention Isolation of suspected or newly purchased birds.

Paratyphoid

The causal agent is a bacterium of *Salmonella* species. It is transmitted by eggs from infected carrier hens. Infected faeces on the egg shell penetrate to the embryo after eight days of incubation. May be transmitted to humans.

Symptoms Young chicks may die without showing symptoms. Other birds lose appetite, show weakness and diarrhoea; they are chilled and huddle for warmth; they lose weight and egg production stops. Mortality is as high as 50 per cent.

Treatment Drugs on prescription. Antibiotics in feed or water may reduce losses. Penicillin and streptomycin in feed or antibiotic/vitamin mixture in drinking water will stimulate appetite.

Prevention Strict hygiene and regular disinfection of the house.

Pullorum (Baccillary white diarrhoea)

Caused by a bacterium of the *Salmonella* species and transmitted through droppings and contaminated equipment.

Symptoms In chicks, acute septicaemia, sudden death; chicks huddle together, and are chilled; they have diarrhoea with white, sticky and foamy droppings; loss of appetite, peculiar 'cheeps' and difficult breathing.

In adults, no outward symptoms in chronic form. In acute form, birds are weak, listless and have greenish-brown diarrhoea. The mortality in chicks may be up to 50 per cent. If infected in the incubator they will die when one or two days old; if in the brooder, they die after one to three weeks.

Treatment Add broad-spectrum antibiotics or sulphur drugs to feed and water to check mortality. Recovered birds should not be used for breeding as they may be carriers.

Prevention Check sanitation and fumigation practices in the incubator. Hygiene: keep brooder and equipment clean.

General diseases

Air sac disease

This is a chronic respiratory disease.

Transmission Outbreaks occur when the birds are in stress, e.g. moving, vaccinations, debeaking, deworming, or when birds are infected by Newcastle disease, bronchitis or coryza.

Symptoms These include difficulty in breathing, loss of appetite and diarrhoea; egg production is reduced to 20–30 per cent. Although mortality is usually low, more culling is required and poor carcasses are found on slaughter.

Treatment Feed tylosine phosphate in feed and water. High levels of antibiotics in feed or water will reduce mortality, will help stimulate appetites and shorten recovery time.

Prevention Water soluble, antibiotic/vitamin mixture in drinking water at times of stress.

Aspergillosis

Referred to as brooder pneumonia, mycotic pneumonia or pneumomycosis. Caused by a fungus.

Symptoms Gasping, croupy breathing, eyes may be inflamed; increased thirst. Mortality is high in young chicks.

Treatment Remove mouldy litter; spray house with 1 per cent copper sulphate; clean and disinfect equipment; treat birds with mycostatin.

Cannibalism

There are a number of factors that may encourage cannibalism:
(a) Overcrowding is a factor so it is necessary to ensure adequate floor space and sufficient allowance for feed hoppers and water containers.
(b) Salt should form 0.2 per cent of the total ration. Also keep the house cool as heat tends to excite the birds and they may turn to cannibalism.
(c) The amount of light allowed into the shed should be kept at a minimum, especially around the nesting boxes.
(d) Supply the birds with enough greenstuff such as cabbage or kale.
(e) Sprinkle grain around to occupy the birds and prevent boredom.
(f) Louse and feather mite infection can lead to cannibalism.

Physical control Part of the upper beak can be cut off to prevent pecking (debeaking) (see Fig. 5.6). Otherwise, use hen blinders, vent shields or anti-peck paste to control pecking.

Coccidiosis

Caused by protozoan of *Eimeria* species. This is a very serious disease of young chicks. It is transmitted from one bird to another through the droppings of infected birds.

Symptoms Bloody droppings, drop in feed consumption, droopy birds, emaciation and high rate of mortality.

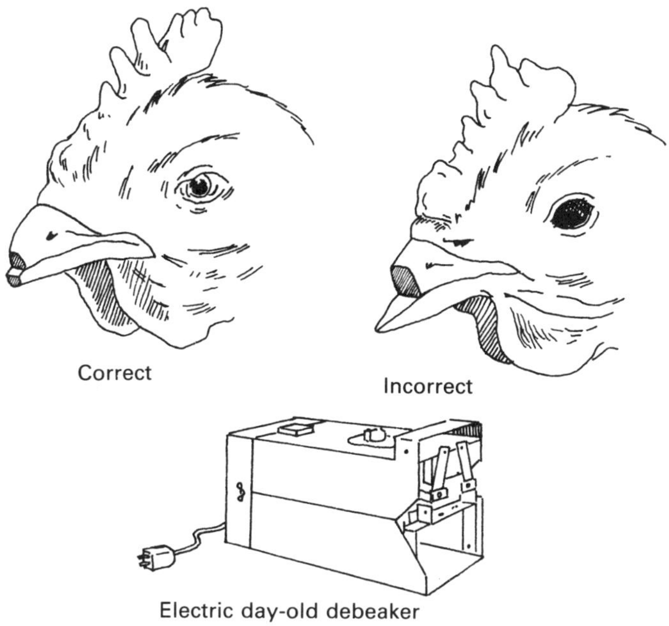

5.6 Debeaking poultry. Only debeak if there is an outbreak of cannibalism.

Treatment Put a chemical such as amprol, sulphamezathine or embazine in the drinking water for seven days.

Prevention Feeding of emprolium, micerbazin or sulfaquinoxaline will prevent outbreaks of coccidiosis. Birds on free range seldom contract this disease. Raise young chicks on wire floor type brooders.

Fowl cholera

It is caused by a bacterium and is transmitted by contamination of feed and water supplies with bowel or nasal discharges from infected birds. Stress conditions (overcrowding, exposure to chills or draughts, unsanitary quarters) weaken the birds' resistance to infection.

Symptoms Dead birds on roosts or floor may be the first signs of the disease. The mortality rate can be as high as 90 per cent. Other signs are profuse, greenish-yellow diarrhoea, loss of appetite, difficult breathing, fever, swollen joints, enlarged comb and wattles, loss of weight.

Chronic carriers have swollen eyes, wattles and facial tissues; also a thick discharge from nasal passages.

Treatment Sulphur drugs in feed and drinking water. Thorough cleaning of quarters and removal of dead and infected birds.

Prevention Maintain high standards of cleanliness and avoid stress.

Fowl typhoid

Symptoms Birds infected may have greenish droppings and loss of appetite.

Prevention Vaccination.

Infectious bronchitis

Caused by a virus and transmitted by air-borne discharge from nose and throat of infected birds. Recovered birds are immune and do not remain carriers.

Symptoms In chicks, a sudden onset and rapid spread of symptoms such as sneezing, coughing mucous and nasal discharge and watery eyes. In adults the same symptoms as in chicks but no nasal discharge. Egg production drops 10–50 per cent. The mortality rate in chicks may be up to 60 per cent and in adults 2 per cent.

Treatment No treatment for virus infection. Prevent secondary infection by adding a mixture of antibiotics and vitamins to water.

Prevention Vaccinate flock.

Infectious coryza (roup)

Caused by a bacterium. Can be transmitted from bird to bird or caused by stress brought on by moving, vaccination, or bad ventilation and overcrowding. It can also be caused through contamination of feed and equipment.

Symptoms Disease may be acute or chronic. Foul smell and discharge from eyes and nostrils, swollen facial tissue and wattles, sneezing, coughing, difficult breathing, drop in feed consumption and egg production. Mortality can be high.

Treatment Sulphur drugs in feed and water will depress symptoms. High level of antibiotics in feed or water will stimulate appetite and reduce danger of secondary infections.

Laryngo-trachetis (chicken influenza)

Transmitted through the air or by contaminated equipment. Recovered birds may be carriers. Caused by a virus.

Symptoms Difficult breathing, coughing, rattling; birds may cough up bloody mucus. The mortality rate may be up to 50 per cent.

Treatment Vaccinate exposed birds not showing symptoms. Stimulate feed intake with wet mash and pellets and use a high-level antibiotic feed. Use an antibiotic/vitamin mixture in drinking water.

Prevention Vaccination of replacement pullets at about six weeks of age in areas where disease is prevalent. Vaccinate broilers at three to four weeks of age. Recovered birds may be carriers and thus should not be used for breeding.

Lymphomatosis (fowl paralysis)

Symptoms Birds rapidly lose condition and die.

Treatment No cure known. Kill affected birds. Rear young stock away from other birds.

Neural leukosis syndrome (Marek's disease)

Cause is not known but may be a virus.

It is highly contagious and transmitted from bird to bird and house to house.

Symptoms Chronic form: symptoms first seen in birds at three to five months. Complete paralysis of body parts, birds lie on their sides with one leg stretched forward and the other one backward. Birds become weak as they do not eat.

Acute form: first sickness shows at two to three weeks of age. Some birds only depressed while others are paralysed; the mortality rate may be 1 per cent per day. The disease may also occur as blindness in one or both eyes.

Treatment Effective treatment is not known.

Prevention Do not mix separate chick broods until after five months. Clean and disinfect houses and equipment, using uncontaminated fresh litter for each new batch. Clean and disinfect areas outside the house, especially paths. Control vermin and wild animals.

Parasites

Parasites may be a major problem if untreated. Table 5.2 lists control measures.

Table 5.2 Parasites of poultry and their control

Name	*Where found*	*Effect on host*	*Treatment*
External parasites			
Lice (Body louse, head louse, shaft louse)	Usually affects adult birds. Lice nits seen on feathers, around vent, breast, under wings and on back	Irritation of skin, scab formation, birds droopy, feathers ruffled, poor weight gains, drop in egg production and increased susceptibility to diseases	(a) Spray or dust birds with a poultry louse compound containing Malathion, Gamma BHC or Carbaryl or other insecticides (b) Mix poultry louse powder into birds' dust bath (c) Rub old engine oil into nesting boxes and perches

Name	*Where found*	*Effect on host*	*Treatment*
Tropical chicken Flea (Stick tight flea)	Cluster on comb, wattles and around eyes	Irritation of eye membrane, may lead to blindness, loss of weight, low egg production or death	As above
Mite (Grey mite and feather mite)	Reddish dark brown spots one eggs, vent, tail, breast of birds; grey mites live on birds at all times	Mites attack feathers, suck blood producing anaemia, loss of weight, reduced egg production, loss of condition, pale comb and wattle	As above
Red mite	Underside of perches, in cracks in woodwork	At night suck blood from chickens	As above
For routine hygiene spray the house, nesting boxes and perches every fortnight			
Internal parasites			
Caecal worm	Blind end of caeca	Transmit black-head organisms; cause unthriftness, emaciation; caeca inflamed and thickened	Phenothiazine
Gape worm (Red in colour and Y-shaped)	Bronchi, trachea, lungs	Pneumonia, gasping, suffocation; high mortality in young birds	Thebensol
Tape worm (Microscopic to 1 cm long; flat white segmented worms)	Small intestines	Loss of appetite, loss of weight, eventual emaciation, diarrhoea; egg production reduced	Follow advice of of veterinary officer

6 Ducks

Introduction

Ducks are common in Asian countries where some farmers may keep just a few birds but there are flocks of several thousand birds. Duck meat and eggs are very popular and command good prices. Ducks generally need less attention and sufer from less diseases than chickens.

Breeding and rearing

Breeds

Local strains of ducks are often crossed with the Khaki Campbell breed. This breed is noted for its egg-laying performance. Peking and Aylesbury breeds are heavier and are good meat producers.

6.1 Duck breeds

Production

Ducks come into lay at about five months old, and good layers will produce 250–280 eggs per year. Ducks will go on laying for a number of years but are usually sold after two years of laying. They will weigh around 1.8 kg at four and a half months.

Rearing

One drake (male duck) is needed for each 25 ducks. A quiet darkened place is required for the nesting place. Muscovy ducks or broody hens make better sitters than most duck breeds. A broody hen should be given at most eight eggs to sit on. A Muscovy duck can manage around 10 eggs.

Most farmers, however, buy their ducklings from commercial hatcheries. The eggs take 28 days to hatch out. Incubators need to be kept at 39.5°C and a high humidity of 80–85% be maintained. Ventilation should be good and the eggs turned three times a day (Fig. 6.3).

Once hatched the young ducklings should be kept in a warm house or yard with the 'mother' for a week, before being allowed out. If the weather is cold or wet they should be kept in for two weeks.

Feeding

Ducks eat fish, fish wastes, kitchen wastes, chopped vegetables, rice bran and similar grain products. Many farmers use duck meal prepared commercially. A meal mixed by a farmer could contain 20 kg broken rice, 12 kg rice bran and 10 kg of fish meal, fish wastes or protein meal. This can be fed mixed with a little water to moisten it.

Water

At all times there must be drinking water available. Ducks like to have water to swim in, but this is not essential.

6.2 Duck house and shade. If there is any danger from predators, cover the front of the house with netting.

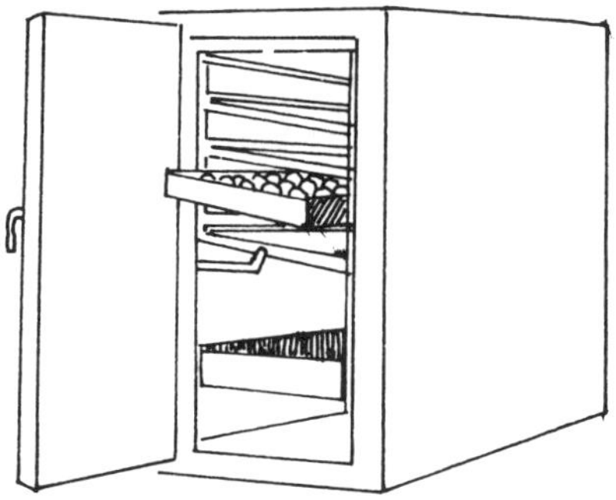

6.3 A duck egg incubator

6.4 Raised duck house over a pond

Housing

Housing can be relatively simple, being of a bamboo and thatch construction. The roof should be rainproof as ducks are not particularly fond of rain and egg production may go down in a rainy period. Ducks lay most of their eggs at night and so eggs should be collected early in the morning to prevent them becoming muddy. Dry nesting places should be provided. There should be shade provided in the duck yard. The yard fence can be of bamboo or chicken wire. It need not be higher than 1 m (Fig. 6.2).

Management of ducks

Ducks are often kept as the only enterprise. However where possible ducks should be included in the farming system. Ducks can be kept in rice fields, providing valuable manure and picking up food in the fields.

Another system is to house ducks by fish ponds and allow them access to the ponds. Their manure is an important food for the fish. They will obtain some food from the pond and can be fed any fish wastes.

If the house is built on poles over the pond it can have a slatted floor so that the droppings can fall through. The ducks will also be safer from predators (Fig. 6.4).

Diseases

Duck plague

This is a fatal disease, caused by a virus found in the faeces. Birds should be vaccinated at two weeks old, have a second vaccination two months later, then a repeat every six months.

Duck virus hepatitis

This disease kills baby ducklings under one week old. The mother should be vaccinated twice before starting to lay, at around four months and then at four and a half months old.

7 Rabbits

Rabbits are raised for meat and can be a quickly grown and are a cheap source of protein. They can be reared for home consumption or as a commercial undertaking.

Breeds often kept are New Zealand White, Flemish Giant and Californian.

Breeding

Breeding stock should be selected from does who are good mothers and who have good body size and shape. The breeding age is five to seven months and the gestation period 30–32 days.

7.1 Rabbit breeds

The doe should be mated again fourteen days after the first mating to ensure a satisfactory service. One buck is sufficient for eight to ten does. The signs of a doe in season are:
(a) she becomes restless;
(b) reproductive organs become swollen;
(c) she will rub against the walls and food containers;
(d) she will throw herself on her side;
(e) she will try to contact the rabbit in the next pen.
Always put the doe into the buck's hutch; never the buck into the doe's hutch as this upsets the buck.

Aim at four litters per doe per year and only allow seven young per litter as more will be too much for the mother. The surplus should be killed after the third day but not before this as the doe will become upset at being disturbed.

Weaning

Weaning should be done at eight weeks; the doe can be mated immediately afterwards.

Feeding

Clean, fresh water should always be available. Provide a piece of rocksalt in the hutch. Feed twice a day, in the early morning and evening. Do not give tomato or Irish potato vines as these are poisonous to rabbits.

Roots Carrots, turnips, sugarbeet and dried cassava chips are suitable.

Green materials Grass, vegetables, sweet potato vines, Napier grass, green maize, kale, green oats, lucerne, clover, cowpeas, soybeans, groundnut vines.

Concentrates Rabbits can do moderately well on greens and roots. However it is best to feed some concentrates. These are pellets or meal containing cereals mixed with a protein feed such as cotton-seed cake, groundnut cake or fish meal.

Management

Nesting

A few days before the doe is due to kindle (give birth) she will go off her food. Provide her with plenty of soft hay for lining her nest box (Fig. 7.2). She will normally add the fur she pulls from her own breast. This will usually be about the 25th day.

Care of the litter

When the litter is born the doe will look thinner, there will be fur and rubbish round the entrance to the nest box, and some

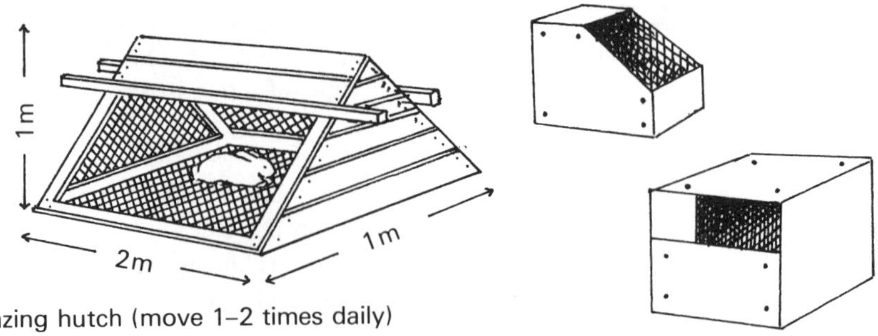

7.2 Grazing hutch and nesting boxes for rabbits

movement will be visible in the dark end of the nest box. Leave the doe alone for at least 24 hours or she may become upset and ignore the litter.

The rabbits are born naked and blind. To inspect the litter, remove the doe first. Remove any dead young and the very weak ones.

During nursing time, offer extra food to the doe in the form of concentrates, dairy cubes, milk and bread.

Weaning

Wean rabbits at eight weeks. Remove the doe to another hutch and leave the young rabbits in the old hutch as this will upset them less.

Records

Records of breeding, feeding and weight gains are important especially for the purpose of selecting breeding stock.

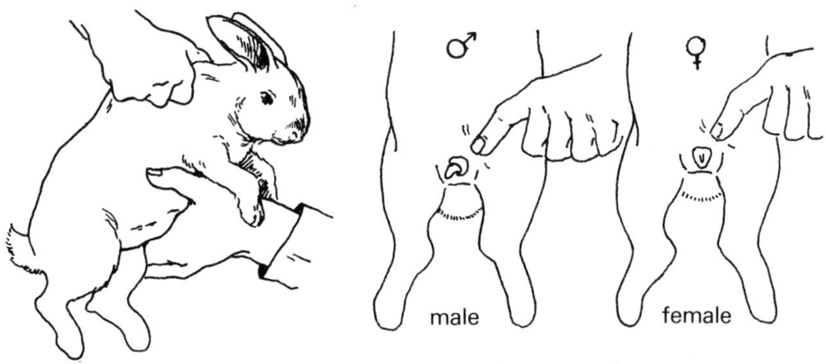

7.3 Handling and sexing rabbits

7.4 Outside rabbit hutch

Housing

Housing should give protection against rain, wind and sun, and also be strong enough to give protection from predators (including dogs and cats).

Hot areas

Build a frame and cover it with wire netting; 2.5 cm mesh wire netting can be used for the top and sides while 1–1.5 cm wire mesh can be used for the floor. The floor can be of timber, but holes must be drilled to provide for drainage. Provide some shade using a thatched roof shelter on poles (Fig. 7.4). If there is danger from predators build a fence around the hutch.

Cooler areas

Make a hutch of timber for the sleeping area and a wire mesh outside run. The hutch can be 120 cm × 60 cm × 60 cm. This is a convenient size and the feed-sunporch area covers half of it.

Grazing hutches

These are movable pens which have a wire netting floor (see Fig. 7.2). They can be 2 m long, 1 m wide at the bottom and 75 cm high. Cover the floor with 3 cm mesh wire, and the sides with 2.5 cm mesh netting. The top of each side can be covered with board or timber, to give some protection from sun and rain. One end can also be timbered for the same reason. These hutches should not be used where there are predators.

Handling

Handle rabbits gently. Lift them by the loose skin at the backs of their necks and support them underneath with a hand (see Fig. 7.3). Do not handle baby rabbits, as the mother may abandon them.

Marketing

Rabbits can be killed for the table at four months of age. For the last month increase the grain feed and decrease the green feed so the rabbits fatten up.

Rabbits skins can be stretched on a board, scraped clean and dried in the shade.

Preventative measures for common diseases

Cleanliness is very important in rabbit keeping. Waterers and feed troughs should be washed and dried in the sun every few days. Wash and disinfect cages when they become empty or each month. Remove manure and dirty bedding every day. Separate any sick animals from the healthy ones and put in a pen away from other rabbits.

Diseases

Coccidiosis

Caused by a protozoan called coccidia. Affects both liver and intestines. Occurs in overcrowded and dirty conditions. Rabbits have swollen stomachs and diarrhoea.

Control Cleanliness; avoid contamination of feed and water by rabbit droppings.

Conjunctivities or weepy eyes

Symptoms Inflammation of eyelids, discharge from the eyes.

Treatment Ointment (e.g. yellow oxide of mercury).

Ear mange or canker

Caused by ear mites.

Symptoms Diseased patches in ears.

Control Dust with dog or cat mange dust.

Pasteurellosis

(Pneumonia, snuffles, septicaemia)

Symptoms Rabbits suffer from sneezing, snuffles and nasal discharge.

Control Add sulphuraquinoxaline to the feed at three or four week intervals. Kill infected rabbits.

Ringworm

Caused by a fungus.

Symptoms Circular patches of scaly skin with red, elevated crusts. Fur falls off.

Control Dust nests with a fungicide.

Skin mange

Caused by mites.

Symptoms Severe itching and scratching. Scaly skin and brown scabs in the ears.

Control Clean all hutches. Treat ears with linseed oil and clean after 12 hours. Apply Terramycin ointment.

8 Sheep

Introduction

In tropical countries the majority of sheep are hardy indigenous breeds. In the higher altitude areas, exotic wool sheep may be kept.

Breeds

Some of the sheep breeds found in the world are listed below and illustrated in Fig. 8.1.

Low rainfall areas

Red Masai

This is a fat-tailed sheep. It has a narrow body and is usually coloured brown, or white and brown. Although slow maturing, it is prolific and hardy, being suitable for semi-arid areas, and can be used for cross-breeding. It is fairly disease resistant. A mature female weighs around 25 kg.

Black Headed Persian

This is a fat-rumped hair sheep with a rather poor quality carcass. However, the sheep are tough and disease resistant and can survive in very harsh conditions.

Somali

These are fat-tailed sheep with white bodies and black heads. They thrive in dry grassland and thorn bush areas. A mature female weighs 35–45 kg.

Kohistani Dumba

This is a fat-tailed, heavy sheep, black or brownish-white in colour with small ears. It is a good mutton sheep and produces some wool (about 5 kg per year). It is found in Pakistan.

Baluchi Dumba

Similar to the Kohistani Dumba but with a better fleece, usually whitish in colour. It is hardier and has floppy ears.

Medium and lower rainfall areas

Merino

This is a hardy wool sheep. It is slow maturing with a light, narrow body. It can walk considerable distances and will survive in the harsher, drier areas. However, it is a poor mutton producer

8.1 Sheep breeds

and is fairly susceptible to foot rot, worms and respiratory diseases. Merinos can be crossed with such breeds as the Dorper to produce a dual purpose meat/wool sheep.

Dorper

Suitable for hot, dry conditions. The Dorper produces a good carcass, but poor quality wool. It originates from a cross between the Dorset Horn and the Black Headed Persian. It is hardy, fairly prolific and has a good growth rate, and is a very useful meat sheep for low and medium potential areas. A mature female weighs 50–65 kg.

Kooka

A thin-tailed good milch breed found in irrigated areas of Pakistan. Polled or with small horns, black in colour, with a roman nose (large curved nose), long neck and with long flat ears.

High rainfall areas

Hampshire Down

These are fertile and early maturing sheep very suitable for cross-breeding for fat lamb production in high potential areas. A liveweight gain of 0.2 kg per day can be achieved.

Romney Marsh

Suitable for high rainfall areas. They are large framed, dual purpose sheep and do well in the high potential areas.

Corriedale

Suitable for medium and high rainfall areas. They have a large body and are suitable for cross-breeding with Hampshire Downs for fat lamb/wool production.

Management systems

Extensive system (arid and semi-arid lands)

Grazing

Sheep are usually grazed in a nomadic, wandering way in the semi-arid and low potential areas. Little attempt is made to undertake controlled or rotational grazing as described for range cattle.

Dipping

The sheep can be dipped one month before lambing to get rid of ticks and parasites.

Footcare

Sheep on soft wet pastures can suffer from lameness as their hooves become overgrown. This can become a serious problem interfering with their grazing, and leading to secondary infections. Regular trimming with a knife is necessary. A disinfectant should be applied where infection occurs or if the trimming results in injury.

Intensive system (good rainfall and pastures)

The following notes are primarily for sheep production in the high potential areas, although good ranchers would follow the same practices as well.

Grazing

In the high potential areas, sheep are grazed on paddocks or may be confined in yards and zero-grazed.

Breeding period

Breeding should take place at such a time that the ewes give birth early in the rains. This will ensure that there is good pasture available for both ewes and lambs.

A ram marker harness can be used to detect unmated ewes. With this harness the ram marks the backs of the ewes he serves. Unlike cows, sheep do not come into heat all the year round but have a definite breeding season.

Selection of breeding ewes

Breeding ewes should be selected on the basis of good size, health, sound teeth and milking ability. In the high potential areas, ewes that produce twins should be retained as this is a valuable characteristic for high production.

Breeding stock should only be purchased from well-managed farms, otherwise there is considerable risk of bringing in parasites and diseases.

Prior to and during the breeding season, rams should be fed well. In the remainder of the year, rams should be kept separate from ewes to prevent unwanted mating. This is not always possible with small flocks, in which case the traditional apron can be attached to the rams (Fig. 8.2).

Each ram can serve 50–60 ewes, but in the harsher range areas, more rams are required (1 to 40) than in the high potential areas.

Flushing

Flushing is the increased feeding of breeding females prior to mating. With sheep this can be done by providing good pasture, fodder crops or a daily ration of 0.5 kg concentrates. The additional feeding should start three weeks before mating and continue for a further three weeks afterwards.

Flushing increases the weight of the ewe and increases her vitality. This helps to ensure successful mating and the birth of a healthy lamb. Ewes should be dewormed two to four weeks prior to lambing, and should be put in a new, clean pasture about one week before lambing.

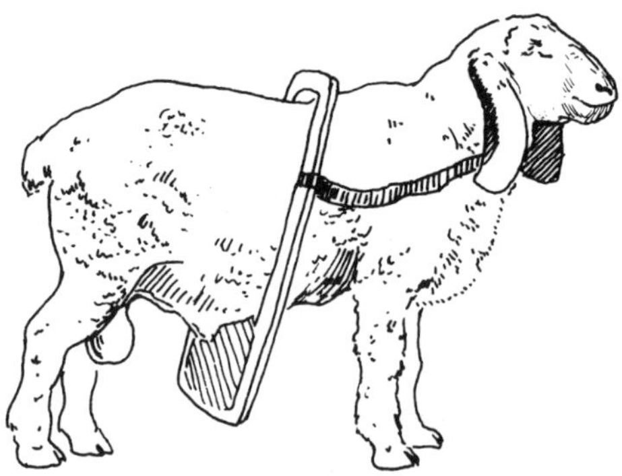

8.2 Apron to prevent ram breeding

Lambing

Pregnant ewes need high quality feeds. If at birth lambs are too small or weak, the losses can be high. However, the ewes should not be overfed so that they become too fat, as this can cause birth difficulties. They can be given a handful of concentrates each day for three to four weeks before and after lambing. No concentrates should be given during the week after lambing otherwise udder problems may develop. Any handling or herding during late pregnancy should be done carefully to avoid stress.

Ewes give birth some two hours after showing the first signs of unease. After lambing, apply tincture of iodine to the navel cord. Lambs suckle within an hour or two of birth, and should be helped if they have not suckled, as losses from hunger and exposure can be high at this time.

A sheltered field should be selected for lambing. In very cold and wet areas, small yards can be made or lambing can take place in a warm, draught-free shed. Ewes with twins will require extra feeding.

Lamb losses occur mostly in the first two weeks and are usually due to: (a) the ewe being excited and not left alone during lambing; (b) a lamb not getting help to suckle when it is weak; (c) the lamb becoming cold and developing pneumonia; (d) dogs and wild animals attacking the sheep. With care and good management most of these losses can be prevented.

Feeding of young lambs

After about six weeks, the milk yield of the ewes declines, so good pasture is even more important for the lambs. By the fourth month, the milk yield is down to 25 per cent. A creep feed area can be made for the lambs where they are fed extra fodder or concentrates.

It is best if the lambs can remain with their mothers all the time and ideally the lamb pens should be near good grazing. However where the mothers have to walk far for grazing, the lambs should not go but remain near the pens.

Castration

Ram lambs not needed for breeding should be castrated at one or two months of age. However, castration is not beneficial in fat lamb production where the lamb will be slaughtered before six months of age. This is applicable to intensive fat lamb production with cross-bred lambs.

Weaning

Weaning can take place at four to five months and the lambs herded together on good pasture for the next six months, separately from adult sheep in order to minimise worm infection. With good management, lambs at weaning should weigh not less than 22 kg.

Shearing

Weaners are sheared at eight months and mature sheep sheared

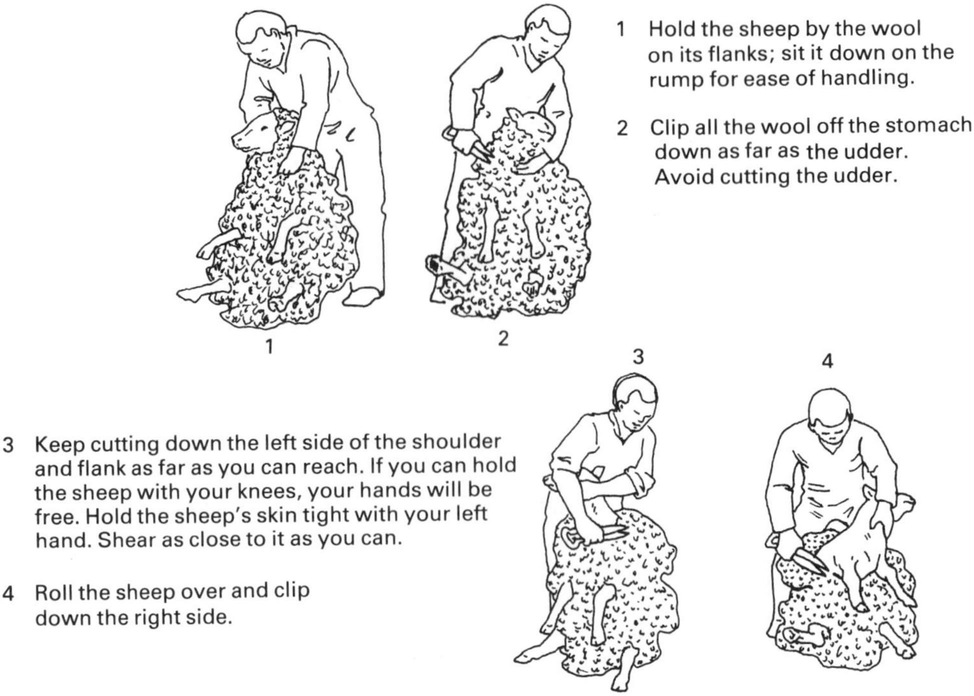

8.3 Sheep shearing

once a year (see Fig. 8.3). Sheep should be dipped six weeks after shearing. Wool sheep also need periodic clipping around the crutch so that they remain clean.

Clipping should be done in dry weather to prevent the newly-clipped sheep suffering from exposure. It also ensures that the clipped fleeces will be stored dry. The sheep should be clipped in a clean place to keep the fleece clean. The clipped fleece can be rolled up to keep it clean and tidy during storage.

Handling

Sheep tend to be nervous and should be handled and driven carefully. Sheep should be caught by the skin on the side of the sheep, not by the wool.

Drenching

Sheep may need monthly drenching during the rains to kill internal parasites such as roundworms. The sheep can be backed into a corner and its head held up. Care must be taken to avoid choking the sheep.

Diseases

Notifiable diseases

The following notifiable diseases of sheep and goats are covered in the section on cattle disease (pages 52–61): anthrax, foot and mouth, heartwater, rabies, trypanosomiasis, tuberculosis.

Sheep pox

This is also notifiable. It is a serious viral disease transmitted by contact with infected animals. The mortality rate can be very high.

Symptoms Small, red pimples around the mouth and under the tail, accompanied by a high fever.

Treatment None.

Control Vaccination.

Other diseases

Blackleg

Symptoms Lameness, fever, swelling of the limbs due to gas formation under the skin. Sometimes occurs after docking or shearing.

Treatment None.

Control Annual vaccination. Careful disposal of carcasses to prevent spread of infection.

Enzootic virus abortion

An infectious disease of sheep producing abortion in late pregnancy.

Symptoms Abortion in late pregnancy, with retained placenta. Sometimes affected lambs are born, alive or dead, at full term. The retained placenta results in uterine infection.

Treatment No treatment to prevent the abortion. Sulphonamides and antibiotics are given to combat secondary uterine infections.

Control Vaccination of susceptible first-lambing ewes and strictly hygienic lambing practices.

Lamb dysentery

Symptoms Sudden death of young at 2–21 days of age, preceeded by diarrhoea and fever.

Treatment Anti-serum to reduce death rate.

Control Vaccination of ewes during last month of pregnancy.

Navel ill (Polyarthritis)

Symptoms Fever, swollen joints.

Treatment Antibiotics are effective in the early stages of the disease.

Control Disinfection of the navel at birth, and of the wounds of castration and ear tagging.

Pulpy kidney

Symptoms Sudden death of the animal, which may occur after a change of diet (as at the beginning of the rains). In an opened carcass, bleeding in the heart and softening of the kidneys may be seen.

Treatment Antibiotics are effective in the early stages of the disease.
Control Vaccination of ewes in the last month of pregnancy and of lambs at weaning. Careful disposal of carcasses by burning or burial to prevent spread of infection.

Table 8.1 Parasites of sheep

Parasite	*Symptoms*	*Treatment*	*Prevention*
Roundworm	Scours, anaemia, wasting	Drench weaners monthly during rains and one month after; see Appendix IV	Rotation grazing; dose ewes just after lambing, and move two days after dosing; rotationally graze and allow lambs access to next new pasture through small gate (This is known as forward creep feeding)
Tapeworm	Wasting, rickets	Drench young stock at six weeks and at weaning;	Rotational grazing; graze young stock first
Liverflukes	Dullness; distended stomach; anaemia	Drench;	Keep stock out of wet pastures and stream banks
Ticks, fleas, lice and scab	Parasites on animals, especially on ears and rump; bare skin with scabby patches.	Dip (see 'Cattle diseases and control')	Dip every week, and more often in high risk areas
Salmonella	Fever; bad-smelling diarrhoea	Antibiotics; see vet	Clean water and feed
Escherichia coli	Watery, yellow diarrhoea; fever	Antibiotics; see vet	Clean housing, water and feed

9 Water Buffalo

Introduction

Water buffalo are mainly located in Asia although there are many buffalo in south-east Europe and the Middle East. The total world buffalo population is currently around 130 million.

Buffaloes have been kept for some four thousand years in China, and almost as long in the Indus valley, Iraq and Iran. In many countries the buffalo is very important from a social and religious point of view. There is considerable buffalo folklore and buffalo may be sacrificed at certain times such as at funerals. They may be used for bride price. There is often great affection between the buffalo and its owner, particularly as young children are often given a buffalo of their own.

Breeds

There are a number of different types of water buffalo with somewhat different characteristics. Thus in India and Pakistan the buffaloes live in river valleys and some of these have been selected for a high milk yield, such as the Murrah, Surti or Kundi breeds. In most of South-east Asia the buffalo are swamp buffalo which are poorer milkers and are used for draught purposes (see Fig. 9.1). Even within countries there are variations in size and colour. In some countries, e.g. Laos, Vietnam, Assam and Sri Lanka, there is interbreeding between wild and domestic buffaloes. In Nepal buffaloes have adapted to living at high, colder altitudes and can be found at 2500 m.

The river buffalo is black or dark grey and often has coiled horns. It is longer, taller and thinner, more refined and prefers clean water to wade in.

The swamp buffalo's skin is grey at birth and later becomes slate blue or slate grey. Whitish animals are not uncommon in Thailand. The swamp buffalo is well built with a large belly, and massive horns that sweep back. Swamp buffalo love muddy water and will wallow for hours in it. Mature female animals may weigh from 350–600 kg, and mature males up to 800 kg.

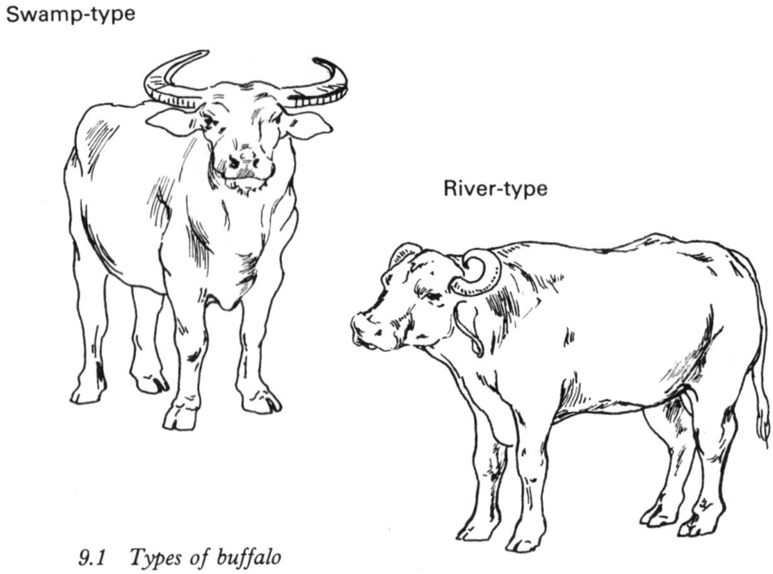

9.1 Types of buffalo

Management

Breeding

The gestation period for swamp buffaloes is around 300–330 days, river buffaloes taking some 310 days. Males can start breeding at around three years of age, and can breed effectively up to eight years of age. One male can service at least 12 cows per year.

Females are normally first bred from around two and a half years and up to three and a half years of age, although they may first come on heat at around 18 months of age. The duration of the heat period is around one and a half days and can be as long as three days. It occurs on a 21-day cycle. The onset of heat for swamp buffaloes often takes place at night and most mating occurs during this time and usually in the cooler season. Calving intervals are normally around 20 months but with careful management this can be improved.

The use of AI is spreading but is comparatively difficult because of the problem of detecting heat and the short irregular periods. Cow buffaloes can be inseminated 24–36 hours after the onset of heat. They can be bred again, when in heat, some 40–60 days after calving. The birth of twins is a fairly unusual circumstance.

Feeding

Buffaloes are ruminants. They eat grass and also browse on leaves and twigs. In addition, in village situations they are given straw to eat. They are able to digest larger quantities of low quality feed, such as wheat and paddy straw, than can cattle. However, animals kept for milk or meat production or worked extensively should be fed additional concentrated feed.

A concentrate ration for milk buffalo can be made up as follows: one part each of cotton seed cake, crushed maize or rice, wheat or rice bran, and grain or some other legume.

Each animal would receive 2 kg per day and, in addition, 1 kg for every 3 kg of milk produced. If available, give 1–2 kg of molasses and a handful of salt per day. Animals normally eat some 50 kg of green feed and 5 kg of straw daily.

Calf management

It is best to leave the calves with their mothers for some months; however, if the buffalo are to be milked then an early weaning is desirable. This can be done at around 40 days of age, but a good concentrate ration will have to be fed to the young calf. Many buffalo calves die because they have not been allowed enough milk as too much has been taken for sale.

When the calves are around six months old after weaning, the males not wanted for breeding can be castrated with a bloodless castrator and also given the necessary vaccinations. (Foot and mouth and haemorrhagic septicaemia).

Milking

River buffalo are sometimes kept for commercial dairy purposes. Often the cream is manufactured into ghee, which keeps for some time and is used for cooking purposes, particularly in India and Pakistan.

The butterfat from buffalo milk is very high and may be around 7–9 per cent, as compared with cows' milk which contains around 3–5 per cent butterfat. For this reason buffalo milk sometimes commands a higher price than cows' milk. With well-managed river buffalo, production may be in the region of 1400–1800 kg of milk in a lactation, although 700–800 kg would be more usual whilst a swamp buffalo is likely to produce only a quarter of this amount. However this could be improved through selection.

A buffalo can milk well for some six or seven lactations, but in fact frequently does for only three or four because of poor management or poor milking techniques.

Meat production

Buffalo meat is comparable to beef. However, the meat from animals at the end of their productive life will never be as good as meat from younger animals reared especially for meat. Animals properly fed can make gains of as much as 0.7 kg per day and should be slaughtered between one and a half and three years of age. Currently, rearing for meat is not very common but there is great potential in a number of countries.

Hides are a valuable by-product, but are often carelessly handled, thus losing value.

Buffalo horn is another by-product and more use could be made of this for ornaments as the horn can be easily worked, and heated and bent. Buffalo hair is used for paint brushes.

Housing

Buffalo are frequently not housed at all. However, some shelter is preferable, especially for milking animals.

Manure is a valuable product. It is often dried and used as a source of fuel when there is a shortage of wood.

Trekking and road work

When trekking, buffalo should not be expected to cover more than 30 km in a day and then not during the hot period of the day.

Buffalo hooves need protection if the buffalo are pulling loads on hard surfaces. The hooves can be shod with metal plates. Sometimes tightly woven straw shoes are tied on or shoes are made from old car tyres.

Handling and care

Buffaloes are gentle, easily handled animals, often looked after by children and women. Rough, noisy handling upsets buffalo and makes them difficult to handle. Buffalo can be easily led but do not like being driven.

Long exposure to sun and high temperature cause very considerable distress as buffalo are not as heat-tolerant as cattle. Buffalo and particularly the young animals, suffer when there is little shade. It is therefore essential to allow the animals access to water to wallow in (see Fig. 9.2) or to hose them down or throw buckets of water over them in the middle of the day.

Dehorning is not often carried out, but could be beneficial where the animals are kept in a feed lot. Dehorning is done in a similar way as for cattle.

Any fencing put up for buffalo has to be strong as they will break normal fences when hungry or attempting to join their herd. Electric fencing has been found to be effective.

9.2 Buffalo wallowing

Cultivation and implements

Buffalo with their large hooves and careful walk are ideal for working in small wet rice fields (Fig. 9.3).

Some of the implements used could probably be considerably improved and made more efficient. There is research work in progress on this. Ploughing and cultivating by mechanical means is several times quicker than with buffalo, and crops can therefore be established more quickly. Using buffalo is probably half as expensive.

A buffalo can plough some 0.2 ha of wet paddy per day, so a farmer rotating two buffaloes can plough nearly half a hectare per day. In dry soil or in heavy clay two buffaloes will be required to pull a single mouldboard plough.

Work periods

Buffaloes are normally only required for heavy work for four or five months of the year when they may work up to five hours per day. They are used to pull ploughs, harrows, levellers, carts and sleds. In addition, they may be used to turn water wheels or walk in a circle around a thrashing floor over a layer of harvested rice in order to trend out the grain.

Buffaloes should only be worked in the morning when it is cooler. Buffaloes like to graze early and again in the evening. In the hot afternoon they prefer to wallow in a muddy pond. This keeps their body temperature down and helps to control parasites.

9.3 Buffalo at work

Training for draught purposes

Training starts at around three years of age and because of the docile nature of the buffalo this is relatively easy. It usually starts with pulling a heavy log around. It is necessary to pierce the nose between the nostrils (nasal septum) and insert a ring or nose peg in order to guide and control the animal.

Buffaloes can continue working for many years, even up to 20 years old; however their normal working life is around 10 years.

Diseases and parasites

Buffalo are liable to contract most of the diseases found in the cattle. The main diseases and parasites are listed below.

Haemorrhagic septicaemia

This is a very serious disease causing many deaths. Preventive vaccinations are necessary, the first at three months old, then at yearly intervals. Pregnant females should not be vaccinated as they may abort.

Foot and mouth disease

As for cattle, buffalo should be vaccinated at six months, then yearly. Pregnant females should not be vaccinated.

Thelazin

This infection attacks the eyes and is caused by worms. It can be treated with a de-worming chemical such as Levamisole.

Liver fluke

This can be treated as for cattle. Infestation is common as the animals are normally in or around swamps. Liver fluke can considerably reduce the productivity of buffalo.

Ascariasis

Buffalo calves are particularly likely to suffer from heavy roundworm infestations which may cause death. Treatment is by dosing with piperazine between two and four weeks of age.

Lice

Spray when lice are present. Clipping hair helps to prevent lice. Wallowing apparently keeps down damage from ticks, flies, mange and lice.

Sources

W. Ross Cockrill FAO *The husbandry and health of the domestic buffalo*

National Academy Press Washington. D.C. *The water buffalo*

Appendix I Glossary

acute form Having violent and rapidly appearing symptoms, leading to a crisis.

anaemia lacking in iron.

antibiotics Chemical substances capable of destroying bacteria or preventing their growth, e.g. penicillin, streptomycin, tetracycline.

antibodies Substances found in the blood that react chemically to destroy invading bacteria and organisms.

artificial insemination The removal of semen from a male animal and placing it into a female's reproductive organs.

bacteria Tiny single-celled organisms; some are capable of causing disease.

boar Male pig.

butterfat The fat element in milk, from which butter is made.

carrying capacity The number of livestock that can be grazed on a pasture.

castration The breaking or cutting of the cords leading to the testicles in the male so that the animal will be unable to breed.

cell The units from which all living organisms are composed.

cellulose The substance making up cell walls and the wood part of a plant.

chronic form Form of disease continuing for a long time.

colostrum The first milky substance to be provided by the mother for her new-born offspring, rich in antibodies and vitamins.

concentrate High-protein feed.

conception State of becoming pregnant; fertilisation of the egg in the female.

cross-breeding Mating different breeds of the same animal together.

crush Structure for holding cattle so that they cannot move (for spraying, etc.).

culling The sale or slaughter of unwanted animals of poor quality.

dormancy A state in which plants, seeds or animals are not growing.

drenching Dosing an animal with a liquid medicine from a bottle.

dry cows Cows not producing milk.

dual purpose Serving two uses.

farrow Sow giving birth.

flaying Stripping the skin or hide off an animal.

fleshing Removal of flesh from the skin or hide.

flushing Feeding additional, high-quality feed before and after service to ensure successful mating and healthy offspring.

foal Young donkey or horse.

fodder Conserved feedstuff, e.g. hay, silage, etc.

foetus The young of an animal in the womb before birth.

freemartin An imperfectly developed female calf incapable of breeding.

free range Not fenced in.

gestation Period between conception and birth, during which the young grows in the mother's body.

gilt A young sow

heat Periodic coming into season of a female animal. ('On heat' is the stage of the cycle when ready for mating).

heifer Young cow which has not yet given birth to a calf.

hormone Substances produced inside the body which control many functions of an animal.

hybrid Offspring produced when two different breeds are mated.

immunity The build-up of resistance to a disease.

in-breeding The breeding of closely related animals.

incubation (a) Period between contact with a disease and appearance of symptoms. (b) The process whereby a bird egg is hatched either by the mother or artificially.

intake The amount of food eaten by an animal.

kid Young goat.

lactation The period of milking from when the cow gives birth to when she dries up.

litter Baby rabbits born at the same time from the same mother.

lesion A damage, an injury.

ley A temporary pasture, sown to grass then ploughed up a few years later.

micro-organism A one-celled animal or plant.

mucous Slimy fluid.

node That part of a stem from which a leaf starts to grow.

oestrus Coming into season or heat in a female.

ovulation The process by which an ovum is released in the female's body.

ovum Female germ cell or egg which after fertilisation develops into a new member of the same species.

pedigree Pure bred and recorded for many generations.

placenta Afterbirth; membrane surrounding unborn young.

poll Hornless.

protozoa One-celled organisms from the animal kingdom; some are capable of causing disease.

quarantine To keep in isolation.

respiration Breathing, lungs.

roughage Fodder containing a high quantity of cellulose or other foodstuff stimulating gut action.

rumen The first and largest stomach of a ruminant animal (e.g. cow).

rumination Chewing of the cud, which is the return of food, previously swallowed, to the mouth for chewing.

scouring Upset stomach, usually in calves, due to a chill or incorrect feeding.

Appendix II Livestock routine health measures

Animal	*Vaccinations*	*Worm treatment* (dose)	*Tick control* (dip)
Cattle:	**Foot and mouth** (a) at four weeks old (b) then every four to six months (c) when outbreak occurs **Brucellosis** (heifer calves only) at three to eight months old **Rift valley fever** (a) at four months old (b) then every year (c) at every three to four months when there is risk of outbreak **Anthrax/Blackquarter** (a) at six months old (b) then every year **Rinderpest** at one year old	(a) at weaning (b) before and after rains	(a) once a week (b) twice a week in high risk areas
Donkeys:	no routine vaccinations	(a) at weaning (b) every six months	only if ticks are seen on animal
Goats:	**Foot and mouth** (a) at four weeks old (b) then every six months (c) when outbreak occurs **Anthrax** at weaning	(a) before mating (b) at weaning time (dam and young)	(a) once a week (b) or twice a week in high risk areas.

Animals	*Vaccinations*	*Worm treatment* (dose)	*Tick control* (dip)
Pigs:	**Anthrax** (a) at five months (b) then every four to six months **Leptospirosis** at six months old	Sows (a) two weeks before farrowing (b) then after weaning Piglets (a) one week after weaning (b) at four months Boars every six months	dust or spray against mites and lice
Poultry:	**Marek's disease** day-old chickens or very soon after **Newcastle disease** (a) at two to three weeks old (b) then every six months **Fowl cholera** (a) at four weeks old (b) then every three months **Fowl pox** (a) at five weeks old (b) then every year **Fowl typhoid** (a) at five to eight weeks old (b) then every three months	no routine measures	none
Sheep:	**Anthrax** at weaning **Pox** (a) at five weeks old (b) then every three years	(a) before mating (b) after weaning dam and young	Wool sheep (a) dip six weeks after shearing (b) dip one month before lambing

Animal	Vaccinations	Worm treatment (dose)	Tick control (dip)
	Blue tongue (a) two months before lambing (b) three weeks before shearing (wool sheep)		
	Clostridial diseases (a) lambs born of vaccinated ewes at four months (b) then every six months		
	Rift valley fever (a) before mating (b) then one month before lambing		

Note: An animal should not be slaughtered for food for three weeks after a vaccination.

Appendix III Animal health products

The following are some of the animal health products currently available which farmers may find useful to keep in hand.

Item	*Product*	*Manufacter/distributor*
Sore and wound treatment	Cooper Healing Oil	Wellcome
	Stockhocm Tar	
	Terramycin Aerosol Spray*	Pfizer
	Terramycin Eye/wound Powder*	Pfizer
	Negasunt	Bayer
	Tetraspray*	Wellcome
	Vioform	Ciba Geigy
Antibiotic/vitamin mixture for calves	Beecham Animal Health	
Branding fluid	Si-Ro-Mark	Wellcome
Bloat remedies	Bloat Guard	Ciba Geigy
	Stop Bloat	Wellcome
Dairy hygiene:		
(a) Udder and milker's hand wash	Anabac	Wellcome
	ICI Udder Wash	Wellcome
	Iosan CCT	Ciba Geigy
	Mastrite	Wellcome
	Orbisan Forte	Beecham Animal Health
(b) Teat dipping	Iosan CCT	Ciba Geigy
	Mastrite	Wellcome
	Orbisan Forte	Beecham Animal Health
(c) Milking utensil cleaner	Bactergent	Wellcome
	Iodron	Wellcome
(d) Milking cream	New Milking Salve	Wellcome

Item	*Product*	*Manufacter/distributor*
Disinfectants	Beloran	Ciba Geigy
	Biocid 30	Pfizer
	Iosan	Ciba Geigy
	Kerol	Wellcome
Fly control	Cooper Fly Spray	Wellcome
	Coopex 25	Wellcome
	Famid	Ciba Geigy
Footrot remedy	Foot Rot Aerosol	Wellcome
	Vioform 140	Ciba Geigy
Lice and manage control	Asuntol	Bayer
	Lindane Dust	
	Neguvon	Bayer
	Poultry Louse Powder	Wellcome
	Sevin Dust	
Milk replacer	Trilk	Wellcome
Mastitis control	(see also Dairy Hygiene)	
	Mastalone*	Pfizer
	Mastitis Test Kit	Ciba Geigy
	Terramycin Mastitis Formula*	Pfizer
Poultry formula	Terramycin Chick and Egg Formula*	Wellcome and Pfizer
Purgative	Epsom Salts	
Rat killer	Racumin	Kleenway/Bayer
	Ratak	Wellcome
	Warfarin 5K	
Scours control	Lectade	Beecham Animal Health
	Terramycin Scours Tablets*	Pfizer
Tick control grease	New Py-grease	Wellcome
Wormers:		
(a) Roundworms	Banminth	Pfizer
	Niverm	Wellcome
	Nilzan	Wellcome
	Panacur	Hoechst
	Rintal	Bayer
	Thibenzole	MSD/Pfizer
	Valbazen	Ciba Geigy

Item	*Product*	*Manufacturer/distributor*
(b) Tapeworms	Mansonil	Bayer
	Nilzan	Wellcome
	Panacur	Hoechst
	Valbazen	Ciba Geigy
(c) Flukes	Bilevon	Bayer
	Nilzan	Wellcome
	Valbazen	Ciba Geigy
(d) Wormers for horses and donkeys	Loxon Horse Wormer	Wellcome
	New Coopane	Wellcome
	Panacur	Hoechst
	Strongid	Pfizer

* Obtainable through a veterinarian.

Index